BIOGRAPH

A Graphical Computer Simulation Package
with Exercises to accompany Lee A. Segel's
Modeling Dynamic Phenomena in Molecular and Cellular Biology

Software

The programs in this book are available on diskettes for IBM-PC personal computers that are equipped with a mathematics co-processor. To get your copy, the cost of which is included in the price of this book, please **tear out** this page, leaving the perforated edge and return it to one of the addresses given below. A copy of the page or a page that is *cut* out will not be accepted as proof of purchase, and no diskette will be supplied if ordered on such a page. There are no cash refunds for diskettes; only diskettes with manufacturing defects may be returned to the publisher.

In the USA and Canada:

Cambridge University Press, Orders Department, 510, North Avenue, New Rochelle, New York 10801.

Elsewhere:

Customer Services Department, Cambridge University Press, Edinburgh Building, Shaftesbury Road, Cambridge CB2 2RU, U.K.

....

Please send me one copy of the **BIOGRAPH** diskette,
ISBN 0 521 33974 X.

Name: ....(Block capitals please)

Address:

....

....

....

....

Signature:

....

BIOGRAPH

A Graphical Computer Simulation Package
with Exercises to accompany Lee A. Segel's
Modeling Dynamic Phenomena in Molecular and Cellular Biology

Garrett M. Odell
Department of Zoölogy, University of Washington

Lee A. Segel
Department of Applied Mathematics,
The Weizmann Institute of Science

The right of the University of Cambridge to print and sell all manner of books was granted by Henry VIII in 1534. The University has printed and published continuously since 1584.

CAMBRIDGE UNIVERSITY PRESS
Cambridge
New York New Rochelle Melbourne Sydney

Published by the Press Syndicate of the University of Cambridge
The Pitt Building, Trumpington Street, Cambridge CB2 1RP
32 East 57th Street, New York, NY 10022, USA
10, Stamford Road, Oakleigh, Melbourne 3166, Australia

First Published 1987

Printed in Great Britain by
Redwood Burn Limited, Trowbridge, Wiltshire

British Library cataloguing in publication data:

Odell, Garrett M.
Biograph: a graphical computer simulation package with exercises: to accompany Lee A. Segel's *Modeling Dynamic Phenomena in Molecular and Cellular Biology*.
1. Biology – Mathematical Models 2. Biology – Data Processing
3. Microcomputers
I. Title II. Segel, Lee A. III. Segel, Lee A. Modeling Dynamic Phenomena in Molecular and Cellular Biology

574'.0724 QH323.5

ISBN 0 521 33973 1
ISBN 0 521 33974 X disc

Contents

Preface

Computer literacy should be a hallmark of every educated person, certainly of every scientist. This manual represents our attempt to integrate computer usage into a course on mathematical modeling for biologists. The contents of this course have been recorded in the text *MODELING DYNAMIC PHENOMENA IN MOLECULAR AND CELLULAR BIOLOGY* (Cambridge, Cambridge University Press, 1984), by Lee A. Segel, hereafter abbreviated MDP or called simply "the text". This manual is a companion to MDP[1].

MDP deals with mathematical models of biological phenomena written as systems of finite difference equations or differential equations. It explains how such models are constructed, then analyzed mathematically. Concepts and qualitative theory are emphasized, not formula manipulation. It gives instruction in phase plane techniques and stability analysis of steady state points. We assume here that readers are familiar with these mathematical concepts.

This manual provides instruction on how to exploit a computer package called "BIOGRAPH " written by G. M. Odell. To get the program diskette, tear out the form at the front of this book (a facsimile is not acceptable) and mail it to Cambridge University Press. The programs that make up the package are written in the "C" language, but the user need not be familiar with this or any other computer language. The package is "menu-driven". Roughly speaking, this means the following. A list of options is displayed on the screen. The user picks one of these by striking a key corresponding to it. A list of sub–options then appears on a secondary menu, and so on. At certain points, questions are asked that require an affirmative or negative response, 'y' or 'n' (e.g. *Change line style?*) or a numerical response (e.g. *Current value of alpha is 0.18975. Enter new value:*).

Suitable menu selections allow the user to obtain solutions for virtually all the systems of equations that appear in MDP. Thus, the numerous graphs of the text can be reproduced. Moreover, it is easy to generate graphs for other parameter values and thereby to sense the full power of the computer to display the consequences of alternative assumptions. Detailed exercises, some with full solutions provided, guide

[1] Errata for MDP (first printing) are given on page 228.

the student in exploiting this power to obtain a deeper understanding of the underlying biology – which is of course the reason for mathematical modeling. Thus, what we present here is not "just another software manual". It is the completion of an integrated educational tool.

The menus provide a variety of flexible options for graphing computed results. In addition to the equation systems in MDP, the program includes several quite general differential and difference equations, whose parameters can be adjusted to model a wide variety of biological phenomena, so that the computer package can even be used as a research tool. Using the program will help the student to acquire two crucial kinds of intuition:

- a geometric understanding of the structure of differential and difference equation models
- a feel for how the dynamical behavior of a differential or difference equation model changes *qualitatively* as a consequence of mere quantitative variation of its parameters.

It can be objected that the package is special purpose, in that it cannot be used to solve equations that might differ only slightly from those in MDP. One reply to this objection is that the limitation of the package to a restricted range of problems is the price that must be paid to make the package work simply and swiftly. This limitation allows the student to concentrate on what the computer can do to enhance biological insight, rather than on programming details of how to make a computer operate. Once a biologist is convinced of the efficacy of the computer as a tool, he or she may be motivated to learn something of programming technicalities. Quite powerful general programs exist whose mastery will not be a problem for a well–motivated student of biology.

This computer package is a *tool* for performing experiments on mathematical models, and this instruction manual is designed in the spirit of a laboratory manual. You will learn more from experimenting than from reading manuals. Thus experiment freely, even recklessly, because even the most careless use of this tool cannot damage it – neither the computer hardware nor software. Scan this manual only briefly[2],

[2]You should probably read at least up to page 26 before you start relying entirely upon experimentation with the program.

and try the program as you read. You should soon have little further use for chapters 1 through 6 of the manual except to clarify subtle aspects that you may discover in the program. You will then be ready to explore Chapter 7, which contains numerous detailed exercises.

Liability disclaimer

We make no warranty, express or implied, that the BIOGRAPH program, or its associated data files and auxiliary programs, are free of error or meet any particular standard of merchantability. Neither do we warrant that BIOGRAPH will meet your requirements for any particular application. The authors and publishers disclaim all liability for direct or consequential damages resulting from your use of this manual and the software supplied with it.

Notation conventions

We use different typefaces in this manual to distinguish facsimiles of text messages displayed on the computer screen from characters that the user types on the keyboard. Here are some examples.

> We will print facsimiles of the messages that appear on your screen in small italic type. The following piece of a menu is an example.

[m]->[m]ake graph from most recent integration data

[q]->[q]uit; clarify screen

[c]->add [c]aption(s) to picture <<F1 MENU>>

[s]->[s]ave picture in a named disk file

[R]->[R]etrieve picture previously saved in a disk file

In this manual, we will print command lines that you type on the computer keyboard in **large bold type**.

We use the ↩ symbol to indicate the carriage return or enter key that you type to end and enter such a command line. The following line is a sample.

alpha ↩

We use single quotes to refer to a key that the user is to strike. Thus '←' refers to the backspace key in the upper right corner of the computer keyboard with ← embossed on it, and 'F5' refers to the function key with F5 embossed on it.

Parts of this manual that describe details, which can be skipped during a first reading, are set in smaller type, indented, and flagged by the ⇝ symbol.

⇝ Here is an example of how such parts are distinguished. We suggest that you should *not* read the material set in this type until you are familiar with how the computer program works. Most of it is included only for completeness. You may find that you *never* need to read it. When you want more information about particular facets of the program, use the index and these small–type segments to find it.

Equation numbers in this manual are written in the style (4.23), while the equation number *(MDP:4.3, p52)* refers to equation 3 in chapter 4 on page 52 in MDP.

References cited in the style of (Smith, 1983) refer to those provided at the end of MDP. References distinguished by an "asterisk", in the format (Jones, 1981)*, are provided at the end of this manual.

Acknowledgements

GMO gratefully acknowledges the support of National Science Foundation grant number MCS 83–01460 for development of the "research version" of the computer program from which BIOGRAPH derives.

Many students who studied MDP and who used early versions of the BIOGRAPH program and manual made useful suggestions for improvements. Of these, the contributions of Volker Brendel and C. Lizarrago were outstanding.

The BIOGRAPH program was created using the excellent DeSmet C development package produced by the C WARE CORPORATION in Sunnyvale California. The authors gratefully acknowledge code optimization assistance from Joel Farley at C WARE.

The manual was created using Donald Knuth's TEX typesetting program with Leslie Lamport's LATEX command macros. See Knuth (1984)* and Lamport (1986)*. Warm thanks go to Carol Weintraub for wise and skilled word processing.

We thank our tolerant and patient editor, David Tranah, for his encouragement and support.

Chapter 1

What BIOGRAPH does and the hardware it needs

1.1 The class of problems addressed

BIOGRAPH is an interactive program that computes numerically, then graphs, solutions to systems of ordinary differential equations (**ODE**'s) and difference equations. Often, only a pair of ODE's is solved:

$$\frac{\mathrm{d}y_1}{\mathrm{d}t} = f_1(y_1, y_2, t) \ , \qquad \frac{\mathrm{d}y_2}{\mathrm{d}t} = f_2(y_1, y_2, t) \ . \tag{1.1}$$

The program is capable, however, of solving the general system of n simultaneous differential equations

$$\begin{pmatrix} \mathrm{d}y_1/\mathrm{d}t \\ \mathrm{d}y_2/\mathrm{d}t \\ \cdot \\ \cdot \\ \cdot \\ \mathrm{d}y_n/\mathrm{d}t \end{pmatrix} = \begin{pmatrix} f_1(y_1, y_2, ..., y_n, t) \\ f_2(y_1, y_2, ..., y_n, t) \\ \cdot \\ \cdot \\ \cdot \\ f_n(y_1, y_2, ..., y_n, t) \end{pmatrix} \ . \tag{1.2}$$

BIOGRAPH can solve a single difference equation

$$y_1[k+1] \ = \ g_1(y_1[k], k) \ , \tag{1.3}$$

or a simultaneous system of n such equations, written in the form:

$$\begin{pmatrix} y_1[k+1] \\ y_2[k+1] \\ \cdot \\ \cdot \\ \cdot \\ y_n[k+1] \end{pmatrix} = \begin{pmatrix} g_1\,(y_1[k], y_2[k], ..., y_n[k], k) \\ g_2\,(y_1[k], y_2[k], ..., y_n[k], k) \\ \cdot \\ \cdot \\ \cdot \\ g_n\,(y_1[k], y_2[k], ..., y_n[k], k) \end{pmatrix} \tag{1.4}$$

We denote the independent variable by (t) in differential equations and we denote the generation number by $[k]$ in difference equations. For j ranging from 1 to n, the $y_j(t)$ are the components of the solution to a differential equation at the time t. The $y_j[k]$ are the components of the solution to a difference equation at generation k.

> Remember our type font convention: parts of this manual that describe details, which can be skipped during a first reading, are set in smaller type, indented, and flagged by the ⇝ symbol.

⇝ If you are familiar with vector notation, you should note that equation (1.2) can be written in the more compact form:

$$\frac{d\vec{y}}{dt} = \vec{f}(\vec{y}(t), t) \ , \tag{1.5}$$

and equation (1.4) can be written as

$$\vec{y}[k+1] = \vec{g}(\vec{y}[k], k) \ . \tag{1.6}$$

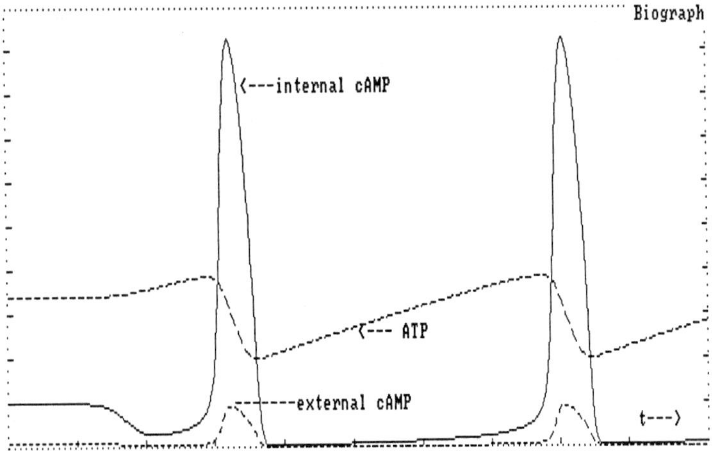

Figure 1.1: cAMP secretion by *Dictyostelium discoideum* amoebae. α =ATP, β =internal cAMP, and γ =external cAMP plotted vs. time t. The plot, a copy of the computer screen made by a printer, shows the 640H $\times$ 200V resolution you will see. The file called camp3osc.pic contains this picture.

1.2 A glimpse of what the program produces

The general differential equations (1.2) and (1.5) may appear forbiddingly abstract, so we give a concrete example of them. The Goldbeter–Segel equations that model the cyclic AMP secretion by *Dictyostelium discoideum* amoebae have three dependent variables, $\alpha(t) = \text{ATP}$, $\beta(t) = \text{internal cAMP}$, and $\gamma(t) = \text{external cAMP}$. These chemical concentrations are governed by the following system of $n = 3$ ODEs:

$$\frac{d\alpha}{dt} = 1 - \sigma\phi(\alpha,\gamma) \ , \tag{1.7}$$

$$\frac{d\beta}{dt} = q\sigma\phi(\alpha,\gamma) - k_t\beta \ , \tag{1.8}$$

$$\frac{d\gamma}{dt} = \frac{k_t\beta}{h} - k\gamma, \tag{1.9}$$

where

$$\phi(\alpha,\gamma) \equiv \frac{\alpha(1+\alpha)(1+\gamma)^2}{L+(1+\alpha)^2(1+\gamma)^2} \ .$$

Equations (1.7)—(1.9) are built into the BIOGRAPH program. They are more fully described on page 33. You can easily produce the plot shown in Figure 1.1 by assigning values to the parameters σ, q, k_t, k, h, and L, determining initial conditions and the length of time to integrate, and selecting a plotting style.

The reason for writing the equations in the general forms, (1.2) and (1.4) is that BIOGRAPH treats any number of equations, n, and it has no way to know special names for the dependent variables such as α. Thus, BIOGRAPH uses the general names, y_1 for the first component, y_2 for the second, etc. This manual and the program itself will tell you how to associate special variable names with y_1, etc. For example, in equations (1.7)—(1.9), y_1 is β, y_2 is α, and y_3 is γ. Thus, in the style of (1.2), equation (1.8) is:

$$\frac{dy_1}{dt} = f_1(y_1,y_2,y_3,t) \equiv q\sigma\phi(y_2,y_3) - k_ty_1 \ .$$

1.3 The style of human–computer interaction

BIOGRAPH is menu-driven: using single keystrokes, you select actions from lists displayed on the computer screen. Many persons prefer command-driven programs to menu-driven programs, because interacting with menus can be slower than initiating commands by typing them. As you become familiar with the program, however, you will memorize the menu choices you use most often. As soon as this starts to happen, the program will acquire the virtues of a command-driven program, while retaining the virtues of menu-driven programs. This is because you don't have to read or point at a menu if you have already memorized the options it offers; you need only strike the single keys that select menu options. These keys have been chosen for their mnemonic value. When you cannot remember what to do next, the menus are always visible and available to help you.

By using the menus and answering queries, you control everything the program does. For example, you control which components are graphed and the plotting style for each. You can obtain traditional graphs of several of the dependent variable components as functions of the independent variable, t, measured along the horizontal axis, as is done in Figure 1.1. Alternatively, you can make phase portraits by plotting one component of the solution *vs.* . another, as illustrated on page 17 in Figure 2.1. You determine how to scale the plots and what kinds of lines to use for each variable – solid lines or arbitrary dot-dash lines. You can partition the screen into one, two, three, or four plot "windows", in each of which several graphs can be drawn.

By making appropriate menu choices you can adjust the values of all the parameters involved in each system of differential or difference equations, such as σ, k_t, and so on in the preceding example.

↝ If you are not acquainted with numerical analysis procedures for solving differential equations, you you can skip this paragraph safely because you can use the program without knowing how the ODE solver algorithm works. Differential equation systems are solved using a variable order (up to sixth), Adams type predictor-corrector algorithm with its integration time step dynamically tuned to ensure that the relative truncation error of the calculation remains below a tolerance you specify. The algorithm is described on page 104. You control the maximum order of the method, the minimum and maximum values of the integration time step allowed, and the number of fixed point iterations allowed for solving the implicit corrector formulas. Any modern text on numerical analysis will discuss the Adams-Bashforth and Adams-Moulton multistep formulae this program uses. See, for example Burden & Faires (1985)*.

By appropriate menu choices, you can save in disk files all the information needed to generate a solution and its plot(s). In particular, "picture" files permit instant screen display of previously generated plots. Other files record parameter values, plotting style, etc. – thereby providing useful aids for reconstructing and modifying previous work. A number of such files comes on your diskette.

1.4 Computer system requirements

The easiest way to see if the BIOGRAPH program will run on your computer system is to turn to the next chapter and try it[1]. If it does not work, this section will help you to understand what your system might need to correct the difficulty.

Be sure to read the file called *readme.txt* on the BIOGRAPH diskette. It contains update information and corrections to this manual. To do this, put the BIOGRAPH diskette in drive a: and type

type readme.txt ↩

or, if you have a printer attached to your computer and wish a printed copy of this information,

copy readme.txt prn: ↩

or

print readme.txt ↩

↝ The program is intended for use on IBM personal computers and close imitators thereof. A minimal hardware configuration is an IBM PC with 256 kilobytes of memory, one double–sided double–density floppy disk drive, a monochrome graphics display adaptor with the standard 200 vertical line × 640 horizontal pixel

[1]See page 8 for detailed step-by-step instructions for novices on how to try the program.

resolution[2], and an 8087 coprocessor chip. The program tests for the presence of these requirements and will not run if they are not met.

The program requires the version 2.0 (or higher) of the PC–DOS (or Microsoft generic MS–DOS) operating system.

The program will run on the IBM PC-XT (with the 8087 coprocessor chip) and the IBM PC-AT[3] The fixed disks in these machines provide much greater file capacity and faster access time to files. The BIOGRAPH program uses disk files extensively and so its performance will be faster on these machines. Response time on the PC-AT is approximately 3.5 times faster than on the PC-XT. But you must use a machine with the 200V ×640H graphics resolution.

The BIOGRAPH program runs successfully on Compaq, Zenith, Tandy, and many other computers that are compatible with the IBM-PC. The program makes no explicit subroutine jumps to BIOS routines. The only hardware feature used explicitly is direct manipulation of the graphics bitmap. Its starting address and storage format must be identical to the IBM PC's. No calls are made to any graphics utility routines your system might or might not have; all graphics primitives are built in.

If you have an IBM graphics printer, an Epson FX–80 printer, an Epson MX–80 printer with the "Graftrax" option, or any other printer capable of graphics screen dumps using the PC–DOS "graphics command", you will be able to print paper copies of any plots you generate. No printer is required, however.

The BIOGRAPH program itself uses 208 kilobytes of memory. In a PC with 256 kilobytes of memory, the remaining memory is available for the MS-DOS operating system and for the occasional loading of the MS-DOS file called *command.com* used to display directories of your disk files on the screen from within the BIOGRAPH program.

The program does no keyboard remapping. It can therefore run while programs such as "Superkey", "Prokey", and "Sidekick" are co–resident in memory and active. The BIOGRAPH program can run under the control of such multi–tasking shell programs as "Desqview" and "Double–Dos". If you use any such

[2]BIOGRAPH will not work with a Hercules graphics card that does not emulate the standard IBM display.

[3]An 80287 math coprocessor chip in the PC–AT plays the role of the 8087 chip in the PC. BIOGRAPH will not run on a PC–AT without the 80287 coprocessor chip.

memory resident programs simultaneously with BIOGRAPH then you must have enough memory to provide BIOGRAPH with at least 208 Kilobytes and to satisfy the memory requirements of those programs and MS-DOS.

When running BIOGRAPH from a floppy disk drive, you can remove the diskette containing the *biograph.exe* file from the disk drive after you have started the program and replace it with a diskette containing data files only.

Chapter 2

How to install and run the BIOGRAPH program

2.1 Simplified instructions for a first session with the program

If you are acquainted with how a PC and its MS-DOS operating system works, you could probably skip to the next section. This section provides detailed step–by–step instructions intended to help those who have never confronted a PC. What you learn by following these instructions should help you to start to make sense out of this manual.

Before you start you will need the BIOGRAPH diskette that comes with this manual[1] and a blank floppy diskette on which to make a copy of it.

The following instructions assume your PC has minimal hardware: 256 kilobytes of memory, only one floppy disk drive, no hard disk drive[2], and no printer. You will need a diskette with the MS-DOS operating system on it. Slip that diskette into floppy disk drive a:[3], with the label on the diskette facing toward the side of the disk drive slot that has the hinged door or lever. Turn the door or lever to lock the diskette in place. Turn on the PC's main power switch if someone else has not yet done so. If the power is already on, hold down the two keys with 'Ctrl' and 'Alt' etched on them and then strike the key labeled 'Del'. These steps cause the MS-DOS operating system to load itself into the

[1] To get the program diskette, tear out the form at the back of this book (a facsimile is not acceptable) and mail it to Cambridge University Press.

[2] If your computer does have a hard disk, turn on the PC's main power switch leaving the hinged doors to all floppy disk drives open and skip to the next paragraph.

[3] if there are two drives use the uppermost or leftmost drive. If the following directions cause a failure, try the other drive.

computer's memory.

It will take some time, possibly more than a minute during which the computer checks for hardware problems, before the computer is ready to interact with you. If asked to supply the date and time, just hit the '↩' key. The MS-DOS prompt should appear on the screen. This prompt should look like "A>" (or "C>" if you have a hard disk).

You should always keep original diskettes containing software you value in a safe place, using them only to make copies that you actually use day–to–day[4]. Floppy diskettes are fragile. In case you lose or destroy the copy, you can make another from the original (archive) diskette. To make a copy of the original BIOGRAPH diskette, issue the following command

diskcopy a: a: ↩

> We use the ↩ symbol to indicate the carriage return or enter key you type to end and enter a command. The preceding line means you type "diskcopy a: a:" and then hit the key with ↩ embossed on it.

Put the MS-DOS system diskette aside and follow the messages displayed on the screen to produce a copy of the BIOGRAPH diskette. In those messages, the "source" diskette refers to the original BIOGRAPH diskette and the "target" diskette refers to the blank one on which you will make a copy.

After the "diskcopy" process terminates, slip the new BIOGRAPH copy into disk drive a: and close the drive's door latch. Give the following two commands

a: ↩

biograph ↩

The program will load and check to see that your PC contains a chip called an 8087 math coprocessor[5] needed to carry out mathemat-

---

[4]BIOGRAPH is not copy protected so that you can follow this maxim. Shun copy protected software that cannot be backed up.

[5]or an 80287 if you have an IBM PC-AT computer

ical calculations rapidly. If that chip isn't installed the BIOGRAPH program won't work. A check is then made to verify that your computer has sufficient memory for the program. Next the program draws a triangle on the screen. This verifies that your PC has standard IBM compatible graphics capability. If you see the triangle, answer the question displayed on the screen by striking the 'y' key. If you don't see the triangle, the BIOGRAPH program will not work on your PC.

Next, a title and copyright message will be displayed in a box on the screen. Hit any key to bring to the screen some text in a box labeled "MAIN MENU". It is the "control center" of the BIOGRAPH program from which you call for other menus that do useful things. What we will do in this preliminary demonstration is call up a file from the disk in which a typical sequence of commands, typed by an experienced user, has been recorded[6]. You will be able to watch the BIOGRAPH program respond to these commands, one by one. To do this, strike the 'F8' function key[7]. This calls up the the command options labeled "F8 MENU".

When you see the F8 menu, strike the upper case 'P' key[8]. The message on the screen, "enter name of playback file (do not type .key)" asks you to type the name of the file that holds a sequence of pre-recorded commands. We are going to use a file called *lesson1.key* with commands that generate graphs of solutions to the simple difference equation (7) on page 17 in MDP

$$y_{i+1} = ry_i(1 - y_i) \tag{2.1}$$

These will be graphs similar to those on pages 22 and 26 in MDP. Respond to the request on the screen by typing[9]

lesson1 $\hookleftarrow$

Read the message in the box in the lower right corner of the screen. Now strike the space bar slowly and repeatedly. Each time you do so,

---

[6]A list of all the files of pre-recorded commands available for similar inspection appears in chapter 6 on page 121.

[7]the key with 'F8' embossed on it

[8]Hold down the key with $\Uparrow$ embossed on it and hit the 'P' key.

[9]You must be precise in typing. You will get an error message if, for example, you add a space after "lesson", or if, for the final character, you hit lower case 'ell' instead of 'one'.

the BIOGRAPH program will respond to the most recent command transferred from the file called *lesson1.key* to the screen. You should read the screen messages, study the commands that are entered automatically from the *lesson1.key* file, and watch how the BIOGRAPH program responds to those commands. Eventually the commands stored in the *lesson1.key* file will be exhausted and you will have to start typing commands yourself. When this happens, either experiment by moving around through the menus or strike the 'F9' function key to terminate the BIOGRAPH program. The narrative below explains some, but not all, of the things the program does while executing the commands from the *lesson1.key* file.

- The 'g' pick from the F7 menu "gets" a file, called *perdoub.par* that selects the general difference equation

$$y[k+1] = ay[k] + by[k]^2 + cy[k]^3 \tag{2.2}$$

  that is described on page 68 in this manual. $y[k]$ denotes the value of the dependent variable at the kth generation.

- The 'i' pick from the F7 menu causes the program to display information identifying the equation currently active (defined by the *perdoub.par* file in this case). Equation (2.2) will be typed on the screen.

- The 'u' pick from the F7 menu initiates a dialog in which the numerical values of parameters in the difference equation (a, b, and c in this case) can be changed. Watch how the values of the parameters a, b, and c in equation (2.2) are set to turn (2.2) into (2.1) with $r = 3.5$

- The next several commands use the options of the F3 menu to determine that:

  1. there will be three windows in the plot and the independent variable (the generation number) will be plotted on the horizontal axis;
  2. the horizontal variable will range from 0 to 75 generations;
  3. the first plot will be drawn in window number 1 (the panel at the bottom of the screen) with the dependent variable ranging from 0.0 to 1.0 .

- The next several commands use the F5 menu to:

    1. determine that the iteration of equation (2.2) is to start at generation 0 (using the 'I' option) and run to generation 75 (using the 'f' option);
    2. specify the initial conditions (using the 'k' option);
    3. generate the first plot (using the 'm' option).

- The next several commands use the 'c' option in the F1 menu to write captions on the plot. These indicate that the plot shows the first 75 generations and they provide the values of the parameter r to be used in each of the three windows.

- The 'u' option of the F7 menu is used to change the value of $a = -b = r$ to 3.65.

- 'F10' is used to return to the main menu from which the 'F3' menu is called

- The 'p' option of the F3 menu is used to schedule the next plot in window number 2.

- The 'm' option of the F5 menu is used to compute and make the second plot.

- The 'p' option of the F3 menu is used to schedule the next plot in window number 3.

- The 'u' option of the F7 menu is used to change the value of $a = -b = r$ to 3.82.

- The 'm' option of the F7 menu calls the F5 menu from which the 'm' option computes and draws the third plot.

- The F8 'S' option (which appears in that menu only while *recording* key files) terminates reading commands from *lesson1.key*.

At this point, hit the space bar twice. If you want to end the session, hit the F9 key to terminate the BIOGRAPH program. Alternatively, hit the F10 function key to return to the main menu and explore the 'tree' of menu options.

This introductory demonstration of what the program does exhibits only a very small subset of the program's capabilities. Nevertheless, understanding this demonstration should pave the way to understanding the rest of the manual. The remainder of this chapter will explain in more detail some of the things you just did.

2.2 How to boot your PC's operating system

"Booting" refers to the process of activating your computer in the fashion of lifting yourself up by your own bootstraps.

If you are using a PC with a hard disk, the operating system files should already be installed on it. You should not need any "system" floppy diskette, and you can turn the computer power switch on and go immediately to the next section. If you have only floppy disk drives, make sure you have version 2.0 (or later) of the MS-DOS (or PC–DOS) operating system. With power to the PC turned off, put the "MS–DOS" operating system diskette in drive a:[10] and close the disk door flap.

Turn on the main power switch of the computer. You will hear the fan motor. After a minute or so, during which the computer checks its circuitry for hardware problems, the light on the disk drive will illuminate and the disk drive will make noise as the operating system loads.

If the PC is already turned on, then do a "warm boot": insert the MS–DOS diskette, hold down both the "Ctrl" and the "Alt" keys, and hit the "Del" key to cause the computer to load the operating system. This warm boot process can be repeated at any time while the BIOGRAPH program is running to start over from scratch. Whenever you do a warm boot, however, you will lose whatever information or graphs you have generated but have not yet saved in disk files.

[10] Usually this is the uppermost or left–most disk drive slot. Treat floppy diskettes tenderly, making sure that unused diskettes are always returned to their protective envelopes. Errors, some unrecoverable, can be caused by dust or liquids, scratches, and deformations of the diskette. Also, be careful not to spill anything on the keyboard.

Enter the date and time if your computer asks for them[11]. Then every file you create will be tagged with the correct date and time, so you will be able to distinguish current and obsolete versions of files[12].

2.3 How to prepare your printer to print plots

With an IBM graphics printer, or an Epson printer (FX-80, FX-85, LQ-1500, or MX-80 with Graftrax), you will be able to print hardcopy graphs of any plot you produce with the BIOGRAPH program. If you don't have such a printer you will nonetheless be able to save your pictures in disk files for later recall and viewing. If you have such a printer, turn its power switch on and prepare your computer to produce plots by giving the following MS-DOS command:

graphics ↩

> One more reminder of our type font convention: in this manual, we will print command lines that you type on the computer keyboard in **large bold type**.
>
> We use the ↩ symbol to indicate the carriage return or enter key you type to end and enter such a command line. "**graphics↩**" above is an example.

After giving that command and producing an interesting plot on the screen, you can get a paper hardcopy of it printed by holding down the shift key and hitting the 'PrtSc' key.

---

[11] The format in which you type a date is, for example,

6-12-1987 ↩

to specify 12 June 1987. Twenty–four hour time format is used to enter times. Thus half past five in the afternoon is entered as

17:30:00 ↩

[12] If you wish you may skip this process by hitting the 'enter' key twice.

2.4 Installing the program

Read the file called *readme.txt* on the BIOGRAPH diskette. It contains update information and corrections to this manual. To do this, put the BIOGRAPH diskette in drive a: and type

type <readme.txt $\hookleftarrow$

Make a copy of the diskette containing the BIOGRAPH program and its related files[13]. Keep the original diskette as an archive or backup version and use only the "working copy" you make. All of the files on your diskette have software protection flags set to prevent their accidental erasure, but they can all be copied to other diskettes. The erasure protection flags are *not* transferred by the copy process.

> $\leadsto$ From within the BIOGRAPH program, you will be able to scan the directories of any active disk drive provided that you have copied the MS-DOS file called *command.com* onto the drive (and subdirectory) from which you run the BIOGRAPH program. Consult your DOS manual for instructions on how to use the "copy" command to copy this file from the system diskette onto the BIOGRAPH diskette. The procedure varies depending on how many disk drives you have. We have left enough free space on the distribution floppy diskette to hold the MS-DOS *command.com* file.

If your computer has only floppy disk drives, remove the MS-DOS operating system diskette from drive a:. Replace it with the diskette onto which you copied the the BIOGRAPH program and associated files, and skip the next section.

Installation in a PC with a fixed disk drive

If your computer is equipped with a fixed disk (hard disk), you should make a sub–directory on it, called, for example, "c:\biograph", then copy all the files from the floppy diskette containing the BIOGRAPH program into that sub–directory. To do this, put the diskette containing

[13] If you have two floppy disk drives, put the MS–DOS system diskette in drive a: and give the command "diskcopy a: b: $\hookleftarrow$", where $\hookleftarrow$ denotes the 'return' or 'enter' key. If you have only one floppy disk drive, give the command: "diskcopy a: a: $\hookleftarrow$". Then follow the instructions displayed on the screen.

the BIOGRAPH program into floppy disk drive a: and then enter this sequence of commands:

c: ↩

mkdir biograph ↩

copy/v a:*.* c:\biograph ↩

copy/v c:\dosdir\command.com c:\biograph ↩

chdir c:\biograph ↩

The expression **dosdir** above represents the name of the sub–directory (or path to it) on your hard disk containing the *command.com* file. You need to copy the *command.com* file onto the hard disk subdirectory only once. The next time you want to use the BIOGRAPH program, just type the following commands and then do the steps in the next section.

c: ↩

chdir c:\biograph ↩

2.5 How to start the program

Whether you are using a floppy disk drive (with the BIOGRAPH diskette in drive a:) or a hard disk and are logged onto that drive's sub–directory containing the *biograph.exe* and *command.com* programs, the next steps are the same. You initiate the BIOGRAPH program by issuing the command:

biograph ↩

You will be asked to verify that the graphics display is working and then you will see a title and copyright notice. Touching any key will bring the main menu to the screen. It indicates the sorts of tasks the BIOGRAPH program does, and looks like this:

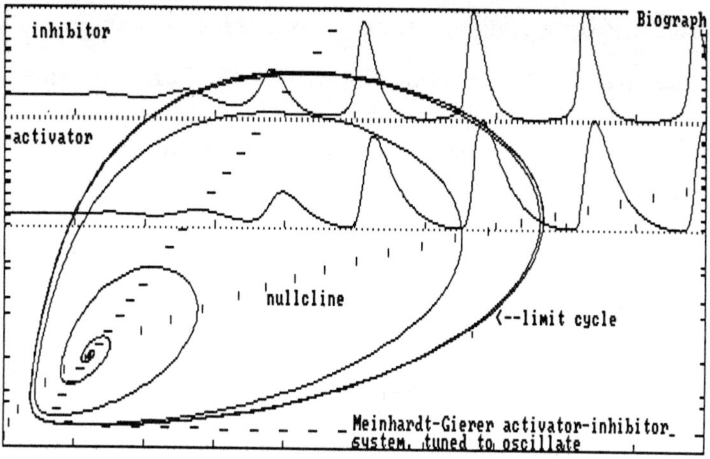

Figure 2.1: A phase portrait for the Gierer–Meinhardt "activator–inhibitor" model. Short horizontal and vertical line segments show nullclines. The two upper panels are windows drawn on top of the phase portrait in which the dependent variables are graphed as functions of time. The file called giermein.pic contains this picture.

F1 = create/clarify/embellish/save/retrieve a picture

F2 = erase the screen

F3 = define plotting style *<<MAIN MENU>>*

F4 = new initial conditions

F5 = integrate/draw solution curves/steady-state points/cursor

F6 = scan data files; print parameter/performance/case data

F7 = identify/get/save the problem-case

F8 = start/stop session log, playback, or keystroke saving

F9 = EXIT PROGRAM

F10 = display this menu

> Recall another type font convention: in this manual, we will print facsimiles of the messages that appear on your screen in small italic type. The preceding menu is an example.

See page 19 for instructions on how to make selections from menus.

Figure 2.1 gives a simple example of one thing that the program does.

Note that if, after aborting the BIOGRAPH program and returning control to the MS-DOS operating system, you wish to re-run BIOGRAPH do not repeat all the steps of this chapter; just re-issue the command:

biograph ↩

2.6 The player piano feature

The BIOGRAPH program can save in a disk file every keystroke you or any user types. That file can be replayed later, after the fashion of a player piano. Playback mode is fully automatic. Without knowing anything about how the program works, you will be able to observe what an experienced user typed to make the program solve some interesting differential and difference equations, and to see what the program did in response to each command typed as was described on page 8. Thus, if you are in a hurry to see the program perform before reading much more of this manual, proceed as follows.

When you see the main menu, strike the F8 key to call up this menu:

– – –{ current path to files = c:\biograph\ }– – –
[L]-> initiate session [L]ogging.
[k]-> initiate disk saving of [k]eystrokes
[P]-> [P]layback a previously saved keystroke file
[d]-> display [d]isk directories (& erase files)
[q]-> [q]uit; continue with program <<F8 menu>>
[!]-> switch to terse menus and prompts
[\]-> change path to files

Hit the 'P' key (note that 'P' is upper case) and you will be prompted to enter the name of a file. Try typing:

giermein ↩

to begin producing Figure 2.1[14]. Alternatively, to begin producing

[14]If you get an error message on the screen, there is something wrong with the

Figure 1.1, type:
camp3osc ↩

Alternatively, to see a sequence of all of the pre-stored pictures, type:
picshow ↩

Next, follow the instructions that appear in the lower right corner of the screen. You have the option of hitting the space bar to initiate each (automatically typed) command after you look at it. That way, you can see what action each command causes. By studying the playback of a few files[15], you will learn quickly how to use the program.[16]

Alternatively, if you just want to "see some action", hitting the '>' key, instead of the space bar, tells the program to execute the commands as fast as it can without waiting for you to hit the space bar. You can restore the halt mode by hitting the '<' key. The *picshow* file played back in fast action mode will display all the pre–stored pictures that come on your diskette.

> ↝ Many auxiliary files are supplied on the distribution diskette. Any file with the '*.key*' suffix can be "played back" as just described. Chapter 6 contains information on how to use those supplied to generate many pictures automatically. Later sections of this manual will explain how to create and edit your own playback files.

2.7 General information on how to respond to menu options and queries from the program

Before explaining how to make the BIOGRAPH program do useful things, we give some general information about the program, in particular about how to recover from possible problems.

---

current "path to files". Use option [d] to change the path. See page 21 for instructions concerning paths to files.

[15]They are identified by the *.key* tag in the file names among the disk files supplied on the diskette. See chapter 6 for a listing of all such files included on the distribution diskette.

[16]See page 8 for detailed step by step instructions for novices on how to try the program.

Special keys

The 'control C' key[17] will abort the BIOGRAPH program. This means you can always make an emergency escape from the program, but no new information you may have wished you stored in disk files will be saved if you exit this way.

Use the 'F9' function key to make a graceful exit. You'll be asked to confirm your intention to quit. Whenever you are prompted to pick an option from a menu, you can, instead, strike the F9 function key, even though that option is not explicitly listed in the menu.

The 'Shift' 'PrtSc' key[18] triggers the computer to reproduce on the printer whatever is on the screen when you strike that key. This works for plots only if, before starting the BIOGRAPH program, you gave the MS–DOS command[19]

graphics ↩

When you are typing input lines in response to queries, the '←' backspace key backs up, cancels, and allows you to re-type characters you typed incorrectly. Note that the "cursor left" arrow, which shares the '4' key on the numerical input pad at the right side of the keyboard, does *not* serve this purpose.

File naming conventions and MS-DOS paths to files

A *file* is a collection of information, stored on a disk, used by the BIOGRAPH program. There are several different kinds of files; they are distinguished by the tag appended to the file's name after the period. The BIOGRAPH program automatically appends the proper tag to file names. Thus, when prompted to type a disk file name, *do not type these tag suffixes*. The following tags are used:

[17] Hold down the 'Ctrl' key while you press the 'C' key.
[18] Hold down the 'Shift' key and strike the 'PrtSc' key.
[19] See page 14.

- **.par** for problem/case parameter files

- **.pic** for screen image (picture) files

- **.log** for session log files

- **.key** for keystroke playback files

- **.ked** for editable versions of *.key* files

- **.exe** for executable program files

The BIOGRAPH program must know where to find your files containing parameter values, stored pictures, etc. Rather than force you to keep such files on a predetermined disk drive, we give you complete freedom to access files stored in any disk drive[20] on your system. At any given time, some particular "path to files" will be active or current, and that path is displayed on the screen as the first line of every menu except for the main menu. When the BIOGRAPH program starts, a certain "default path" is selected. You can change this path at any time from any menu, except for the main menu, by selecting the [\] option. If you make this [\] choice from any menu, the following message will

[20] or any subdirectory of a hard or soft disk if you partition your disk drives using the tree–structured file system provided by the MS-DOS operating system

appear.

Enter new path to files

To use the default path just enter

:

To use the currently active subdirectory of drive b, for example, enter

b:

To use the root directory of drive c, for example, enter

c:

To specify a particular subdirectory, enter the entire path string terminated with '\'. For example

c:\biograph\odellgm\picsafe

current path = default path

*Enter * to keep present path*

Enter the new path here - - >

The "default path", selected by entering

: ↩

refers to the drive/subdirectory from which you gave the following command to start the program.

biograph ↩

For example, suppose that you are using the BIOGRAPH diskette in drive a: and wish to store a new case file on a disk in drive b:. This is done by hitting the 'F7' key, then the 'S' key — but first you must hit the '\' key and type

b: ↩

to specify that the file is to be stored on drive b:.

↝ To give a more complex example, suppose your copy of the *biograph.exe* program and *command.com* program reside in subdirectory c:\biograph\. Suppose, after booting your system, you gave this sequence of commands.

b: ↩

chdir b:\project1\params ↩

c: ↩

chdir c:\biograph ↩

biograph ↩

Then, at the time you initiated the BIOGRAPH program,
b:\project1\params\
was the active subdirectory of drive b:, and c:\biograph was the active subdirectory of drive c:. Thus if, in response to the above screen message, you enter either

: ↩

or

c: ↩

or

c:\biograph ↩

then the program's path to files will become c:\biograph\. If you enter

b: ↩

or

b:\project1\params ↩

then the program's path to files will become b:\project1\params\. If you enter

c: ↩

then the program's path to files will become the root directory of drive c:

As an example of the use of this feature, instructor Brown could request students Smith and Jones to keep their files in subdirectories *c:\brown\smith* and *c:\brown\jones* respectively. This makes it easy to partition different students' files to inspect, grade, and erase them at the end of a course.

Typing numerical input

Whenever you initiate a step that calls for numerical input, *you must enter some number*. If you do not want to enter a new number, try to enter the "present value", which the program usually displays before asking you for a new value. Some numbers are inherently integer quantities, for example the number of "windows" to be used in a plot. Enter them without any decimal point. Ordinary decimal or "floating point numbers", such as −.0000123, can be entered in either of two principal ways:

−.0000123 ↩

or

−1.23E−5 ↩

−1.23e−5 ↩

To enter a floating point number such as 6.00, the decimal point is not required. That is, you can just type

6 ↩

Typing text input

Whenever you initiate a process of text input, such as caption text, *you must enter some text*, technically called a "text string". If you wish to abort the process of text input, enter a junk string, such as a blank caption. Whenever you are asked to type the name of a file to read or write, you can abort the operation by entering

* ↩

If you are asked a question to which the answer is "yes" or "no", type 'y' or 'n'[21].

[21] Actually, *any character* other than a 'y' or 'Y' is interpreted as a "no" answer.

What to do when lost

If you get lost amongst the menus, there are several things to try.

- Hit the 'F10' function key to invoke the main menu.
- Hit the 'q' key (to quit the present process and return to the previous menu).
- Hit the 'space' key and then the '↵' enter key. The space will interrupt the cursor move procedure or the caption write procedure. The enter key will enter a [0.0] number if the program is waiting for entry of a number, and will enter a null string if the program is awaiting text input.
- During the cursor movement procedure, the 'h' key will redisplay instructions for moving the cursor.
- Hit the F9 key to make a proper exit from the program.

If you *really* feel lost and wish to get back to the MS–DOS operating system to start the whole process over from the very beginning (e.g. if the program dies completely and will not show you any menus or respond at all), there are two possible recovery methods. One is the warm boot (insert the MS-DOS diskette and hit the Ctrl-Alt-Del key combination) described on page 13.

If even this does not work, perform a cold start. Turn off the PC's main power switch. Put the system diskette in drive a: if your system uses only floppy disk drives. Turn the computer's power switch back on[22] and boot the system again. Note that some PC's have a special button on the back to force a cold reboot, without turning the power switch off then on again.

We have tried to construct the BIOGRAPH program so that it never "dies" or becomes unresponsive, so you should never need to resort to the drastic measures in the last two paragraphs. If you do, please write

[22]If you are using a PC with a hard disk, open the floppy disk drive doors before turning the power back on. Some PC's have built in circuitry to sense power glitches and automatically disconnect the power supply from the power company so the computer won't get damaged. If you have such a PC, you may have to wait a full minute before turning the computer back on, because it can take that long for this guard circuitry to "relax" after a power interruption.

down a detailed description of the difficulty you had and send it to one of the authors. We will try to fix the problem.

Chapter 3

Differential equation problems BIOGRAPH can solve.

3.1 General structure of the BIOGRAPH program

The program called *biograph.exe* on your diskette was made by configuring general purpose, interactive, numerical ode solver, and computer graphics display algorithms into code specialized to solve about twenty–four difference and ordinary differential equation systems. Most of these equations are discussed in MDP, but a few venture further. Code segments were written to evaluate the right hand sides of each of these equations and to make a user–friendly menu–driven mechanism for changing the numerical values of the parameters appearing in each case. Thus, you do not need to know the C computer programming language, in which the BIOGRAPH program is written, to have a C compiler for your personal computer, or to write any code, to do the exercises the manual describes. One drawback is that you cannot use the BIOGRAPH program to study *other* ode systems in which you have an interest. Nevertheless, we supply a second order system whose right hand sides are generic cubic polynomials and a third order system whose right hand sides are generic quadratic polynomials plus selected cubic terms. These systems are quite general. At a future date, we will publish a version of the program having a built–in compiler for a simple high level language for specifying arbitrary new right hand sides.

3.2 Specific ODE problems and their cases

We distinguish equation systems by calling them "chemostat problem", "cAMP relay–oscillator problem", and so on. Each of these problems can have many different cases corresponding to different parameter values, different styles of graphing solutions, etc. Information defining each case is stored in a disk file identified with a "*.par*" tag. For example, the file called *enz_sub.par* defines one case of the third order enzyme–substrate system consisting of equations (13a) and (13b) on page 54 of MDP and equation (7d) on page 53. While running the BIOGRAPH program, you pick one of these problems (*and you must pick one*) by reading a case file from the disk. In addition to the parameters in the system, each case has a special parameter, called *select*. For the time being, you can think of this parameter as simply an identifying number for problems, not to be changed except under special circumstances.

> ↝ Legal values of the *select* parameter are 1—13, 19, 20 and 21—29. *Select* between 1 and 20, specifies differential equation problems. Values of *select* > 20 specify finite difference equation systems. The part of the computer program code that evaluates the right hand sides of the equation system uses the value of the *select* parameter to decide which system of equations to compute. If you alter the value of the *select* parameter, you pick another equation system but you will continue to use the parameter names and values that match the previous system. Except for certain special cases, to be described below, that would lead to nonsense calculations. Thus do not change the value of the parameter called *select* except for problems for which that change is suggested explicitly below.

Note that, using the BIOGRAPH menus to be described in chapter 5, you can change completely the defining properties of a case file. After discovering a "good" set of parameter values and an appropriate style for graphing the solutions, you will be able to write a new case file on the disk, giving it any name you please. In this way you can assemble a library of case files, each of which will cause the BIOGRAPH program to make a "nice" plot. We have supplied some as templates for you to modify. These are "*∗.par*" files where ∗ stands for a particular file name, usually mnemonic in character.

The rest of this chapter comprises a list of the differential equation problems you can study. For each we supply a table specifying the text names you will type to refer to the parameters occurring in the book (many of which must be slightly modified since they are un–typeable Greek letters or use subscripts). We also list associated files supplied on your diskette. These may be the **.par* case template files, **.pic* picture storage files, or **.key* step–by–step "player piano" files that provide illustrations of suitable calculations and graphics (see page 18 and see Chapter 6 for a listing of what the **.key* files supplied do).

The program distinguishes upper and lower case letters in menu choices and parameter names. Thus 'theta', 'Theta', and 'THETA' are all distinct names.

Each parameter has an ID (Identification) number and that number also appears in the table that is given with each problem below. We recommend, however, that beginners ignore these parameter ID numbers completely.

> ↝ Inside the BIOGRAPH program, only the parameter ID numbers are important. Later when you are an expert, you can change the parameter text names to anything you like in the **.par* files. For several generic equation systems (e.g. the *select* = 5 and *select* = 26 problems), we have used general parameter names, such as a_{12}, a_{20}, etc., which have no mnemonic value. When you specialize these generic systems for a specific application, it is appropriate to introduce meaningful names for the parameters. To do so, use any text editor program to modify the **.par* file, and substitute the names of your choice in place of the default names. In the **.par* files, only lines beginning with the characters '@P' contain parameter name and value specifications. While editing, preserve the spacing and alignment of the "=" signs in the **.par* file, and make sure that your text editor does not insert into the **.par* files anything except standard alphanumeric characters. In particular, make sure your text editor does not contaminate the **.par* files with control characters. If you eliminate a parameter by deleting its name and ID number from a *.par* file, that parameter will be assigned the value 0.0 when the edited *.par* file is read.

Equation numbers in this book are written in the style (3.2), while we use the following style to cite equations in MDP. *(MDP:4.3, p52)* refers to equation 3 in chapter 4 on page 52 of MDP. Sometimes we omit the page references. The symbol $\aleph 12_6$ refers to an exercise in chapter 7 of this manual, namely exercise number 12 associated with chapter 6 in MDP.

3.3 Chemostat problem

Select = 1 identifies the second order chemostat problem consisting of equations *(MDP:5.21a)* and *(MDP:5.21b, p84)*[1], modified by the addition of one parameter, γ. The equations are:

$$\frac{dn}{dt} = \frac{\mu nc}{1+c} - n \tag{3.1}$$

$$\frac{dc}{dt} = \xi - \gamma c - \nu\frac{\mu nc}{1+c} \tag{3.2}$$

When $\gamma = 1$, equation (3.2) is identical to equation *(MDP:5.21b)*. General values of γ are appropriate for a predator–prey interaction in a chemostat; see exercise $\aleph 5_5$ on page 150. Component 1 is n and component 2 is c. That is, in the nomenclature of equation (1.2), $y_1 = c$ and $y_2 = n$.

parameter in manual	computer name	ID number
μ	mu	1
ξ	ksi	2
ν	nu	3
γ	gamma	4
	select	39

Associated files supplied		
.par	.pic	.key
chemo		

Remember that the beginner can completely ignore the ID numbers.

NOTES[2]

[1]This notation refers to equation number 5.21b on page 84 in MDP.

[2]We begin the description of each separate problem on a new page. You may find the blank space thus created convenient for writing notes.

3.4 cAMP relay–oscillator problem: second order approximation

Select = 2 identifies the second order cAMP relay–oscillator problem consisting of equations *(MDP:6.15a)* and *(MDP:6.15b, p111)*, augmented with two new parameters, s and k'. Non–zero values of s and k' give an interesting generalization of the original model. The equations are:

$$\frac{d\alpha}{d\tau} = 1 - k'\alpha - \sigma\phi\left(\alpha, \frac{k_t\beta}{hk} + \frac{s}{k}\right) \tag{3.3}$$

$$\frac{d\beta}{d\tau} = q\sigma\phi\left(\alpha, \frac{k_t\beta}{hk} + \frac{s}{k}\right) - k_t\beta \tag{3.4}$$

where

$$\phi(\alpha, \gamma) \equiv \frac{\alpha(1+\alpha)(1+\gamma)^2}{L + (1+\alpha)^2(1+\gamma)^2} \tag{3.5}$$

Component 1 is β and component 2 is α. Even though this is a second order system, we have added a third dependent variable, y_3, that facilitates compatibility with the full third order cAMP problem listed next. y_3 is γ described in the next (*select* = 3) case, and the differential equation for it just keeps $\gamma(t)$ equal to its initial value. By alternating the value of the *select* parameter between 2 and 3, you can switch back and forth between the full third order system and this second order approximation of it [3].

↝ The addition of a third differential equation to this case to facilitate compatibility with the *select* = 3 case has the following side effect. When you use the F5≻[c] option to find steady–state points and eigenvalues of the linearized system, the eigenvalue −1.0 will be reported. This is the eigenvalue associated with the (fake) differential equation for $y_3 = \gamma$. It is *not* an eigenvalue of (3.4) and (3.3).

---

[3] That is, use the [u] choice in the F7 menu to change the value of the parameter called *select* to either 2 or 3.

parameter in manual	computer name	ID number
σ	sigma	1
q	q	2
k_t	kt	3
k	k	4
h	h	5
L	L	6
k'	kprime	7
s	s	8
	select	39

Associated files supplied		
.par	.pic	.key
2d_camp	2d_camp	2d_camp

↝ Note: there are errors in Figure 6.15 and its caption text in MDP, caused by blunders in shifting from the dimensional variables of Figure 1 of Goldbeter and Segel (1980) to dimensionless variables. 1.4 should be subtracted from the coordinates given on both the ordinate and abscissa of Figure 6.15. k_t should be 10.0, and k' should be $k' = k_1 k S/V$.

See the *select* = 9 problem on page 48 for a similar system of equations modeling glycolytic oscillations.

NOTES

3.5 cAMP relay–oscillator problem: full third order system

$Select = 3$ identifies the third order cAMP relay–oscillator problem consisting of equations *(MDP:6.3a)*, *(MDP:6.3b)*, and *(MDP:6.3c, p103)*, augmented by the addition of two new parameters, s and k'. See page 212 for a full derivation of equations *(MDP:6.1a)—(MDP:6.1c, p102)*, of which the following equations are dimensionless versions. The equations are:

$$\frac{\mathrm{d}\alpha}{\mathrm{d}\tau} = 1 - k'\alpha - \sigma\phi(\alpha,\gamma) \tag{3.6}$$

$$\frac{\mathrm{d}\beta}{\mathrm{d}\tau} = q\sigma\phi(\alpha,\gamma) - k_t\beta \tag{3.7}$$

$$\frac{\mathrm{d}\gamma}{\mathrm{d}\tau} = \frac{k_t\beta}{h} - k\gamma + s \tag{3.8}$$

where

$$\phi(\alpha,\gamma) \equiv \frac{\alpha(1+\alpha)(1+\gamma)^2}{L+(1+\alpha)^2(1+\gamma)^2} \tag{3.9}$$

Component 1 is β, component 2 is α, and component 3 is γ. When $k' = s = 0$, these equations reduce to the system *(MDP:6.3, p103)*. Note that the parameters are the same in this problem and the preceding one. You can toggle (*i.e.* switch back and forth) between the second and third order versions of the cAMP relay–oscillator model by alternating between $select = 2$ and $select = 3$, without reading in a new *.par* file[4].

Figure 1.1 on page 2 shows one of the figures produced by the *camp3osc.par* and *camp3osc.key* files. See the $select = 10$ problem for a generalization of these equations that exhibits deterministic chaos.

---

[4]That is, use the [u] choice in the F7 menu to change the value of the parameter called *select* to either 2 or 3.

parameter in manual	computer name	ID number
σ	sigma	1
q	q	2
k_t	kt	3
k	k	4
h	h	5
L	L	6
k'	kprime	7
s	s	8
	select	39

Associated files supplied		
.par	.pic	.key
camp3osc	camp3osc	camp3osc

NOTES

3.6 cAMP relay–oscillator problem: alternate 2nd order approximation

Select = 13 identifies a quasi–steady–state approximation to the full third order cAMP relay–oscillator system given in the preceding case. This system is an alternative to *(MDP:6.15a)* and *(MDP:6.15b, p111)*. It is a second order system derived by assuming that the intracellular cAMP concentration, β, is in quasi–steady–state equilibrium. When (3.7) is set equal to zero, we obtain:

$$\beta = q\sigma\phi(\alpha,\gamma)/k_t \tag{3.10}$$

This value for β substituted into (3.6) and (3.8) yields

$$\frac{d\gamma}{d\tau} = \frac{q\sigma}{h}\phi(\alpha,\gamma) - k\gamma + s \tag{3.11}$$

$$\frac{d\alpha}{d\tau} = 1 - k'\alpha - \sigma\phi(\alpha,\gamma) \tag{3.12}$$

where

$$\phi(\alpha,\gamma) \equiv \frac{\alpha(1+\alpha)(1+\gamma)^2}{L+(1+\alpha)^2(1+\gamma)^2} \tag{3.13}$$

Exercises $\aleph 14_6$, $\aleph 15_6$, and $\aleph 16_6$ involve this system. Note that you *cannot* switch between this case and the *select* = 2 and *select* = 3 cases by changing the value of the *select* parameter.

Component 1 is γ and component 2 is α.

parameter in manual	computer name	ID number
σ	sigma	1
q	q	2
k_t	kt	3
k	k	4
h	h	5
L	L	6
k'	kprime	7
s	s	8
	select	39

Associated files supplied		
.par	.pic	.key
gamacamp		

3.7 Liver regeneration problem

Select = 4 identifies the second order liver regeneration example consisting of equations *(MDP:A5.19, p259)*.The equations are:

$$\frac{\mathrm{d}v}{\mathrm{d}t} = v\left(\delta e^{-s} - \rho\right) \tag{3.14}$$

$$\frac{\mathrm{d}s}{\mathrm{d}t} = \frac{\gamma v}{1+v} - s \tag{3.15}$$

Component 1 is v and component 2 is s.

parameter in manual	computer name	ID number
δ	delta	1
ρ	rho	2
γ	gamma	3
	select	39

Associated files supplied		
.par	.pic	.key
liver		liver

NOTES

3.8 Generic second order system with cubic polynomial right hand sides

select = 5 identifies a generic second order system with cubic polynomial right hand sides. This system does not appear in MDP, but many equations therein are special cases of it. The equations have this form:

$$\frac{dx}{dt} = \sum_{i=0}^{3} \sum_{j=0}^{3-i} a_{ij} x^i y^j \tag{3.16}$$

$$\frac{dy}{dt} = \sum_{i=0}^{3} \sum_{j=0}^{3-i} b_{ij} x^i y^j \tag{3.17}$$

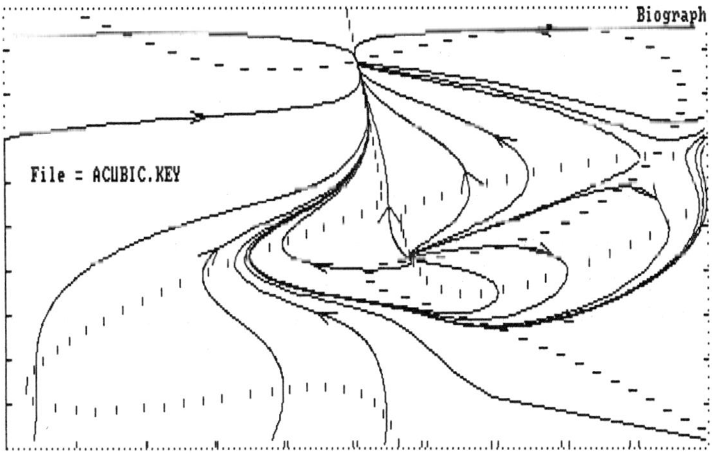

Figure 3.1: A sample second order system with cubic right hand sides. Its phase portrait has three steady–state points. One branch of the vertical nullcline forms a closed curve. The values of the parameters used to generate this figure can be found in the file acubic.par, and the file acubic.key generates the picture.

The summation notation in (3.16)–(3.17) is a condensation of:

$$\frac{dx}{dt} = a_{00} + a_{01}y + a_{02}y^2 + a_{03}y^3 + a_{10}x + a_{11}xy$$

$$+a_{12}xy^2 + a_{20}x^2 + a_{21}x^2y + a_{30}x^3 ,$$

$$\frac{dy}{dt} = b_{00} + b_{01}y + b_{02}y^2 + b_{03}y^3 + b_{10}x + b_{11}xy$$

$$+b_{12}xy^2 + b_{20}x^2 + b_{21}x^2y + b_{30}x^3 .$$

Component 1 is x and component 2 is y. The following assignment of parameters generates the competing species exercise number 10 on page 95, in MDP. $N_1 = x$, $N_2 = y$, all a_{ij} and $b_{ij} = 0$ except

$$\begin{array}{lll} a_{10} = r_1 - q, & a_{12} = -r_1\alpha_{12}/k_1, & a_{20} = -r_1/k_1, \\ b_{01} = r_2 - q, & b_{12} = -r_2\alpha_{21}/k_2, & b_{02} = -r_2/k_2, \end{array} .$$

The file called *figu.par* is used in exercise $\aleph 2_{A5}$ on page 195.

The parameter assignment made by the "acubic.*" files gives rise to the phase portrait shown in Figure 3.1. If you use the player piano feature to play back the *acubic.key* file, you can learn most of the operations you need to construct such a complex phase portrait.

parameter in manual	computer name	ID number
a_{00}	a00	1
a_{10}	a10	2
a_{01}	a01	3
a_{11}	a11	4
a_{20}	a20	5
a_{02}	a02	6
a_{21}	a21	7
a_{12}	a12	8
a_{30}	a30	9
a_{03}	a03	10
b_{00}	b00	11
b_{10}	b10	12
b_{01}	b01	13
b_{11}	b11	14
b_{20}	b20	15
b_{02}	b02	16
b_{21}	b21	17
b_{12}	b12	18
b_{30}	b30	19
b_{03}	b03	20
	select	39

Associated files supplied		
.par	.pic	.key
acubic figu		acubic

NOTES

3.9 Gierer–Meinhardt activator–inhibitor problem (spatially homogeneous version)

Select = 6 identifies the second order activator–inhibitor problem consisting of equations *(MDP:8.1a)* and *(MDP:8.1b, p150)*, with D_a and D_h both set to zero. The equations are:

$$\frac{da}{dt} = \rho_a \left(c_a + c\frac{a^2}{h} \right) - \mu_a a \tag{3.18}$$

$$\frac{dh}{dt} = \rho_h c_h a^2 - \mu_h h \tag{3.19}$$

The *select* = 20 case on page 56 provides a simulation that keeps the diffusion terms, here set to zero, to model reaction–diffusion pattern formation in a ring of cells. This (*select* = 6) case can be used to study the behavior of the morphogen biochemistry that each cell in the ring individually carries out.

Component 1 is h, the "inhibitor", and component 2 is a, the "activator". Figure 2.1 shows what the *giermein.key* file produces.

parameter in manual	computer name	ID number
c_a	ca	1
c	c	2
μ_a	mua	3
c_h	ch	5
μ_h	muh	6
ρ_a	rhoa	8
ρ_h	rhoh	10
	select	39

Associated files supplied		
.par	.pic	.key
giermein	giermein	giermein

3.10 Third order enzyme–substrate problem

Select = 7 identifies the third order enzyme–substrate problem consisting of equations *(MDP:4.13a)* and *(MDP:4.13b, p54)*, with *(MDP:4.7d, p53)*. The equations are:

$$\frac{dS}{dt} = -k_1(E_0 + C_0 - C)S + k_{-1}C \tag{3.20}$$

$$\frac{dC}{dt} = k_1(E_0 + C_0 - C)S - (k_{-1} + k_2)C \tag{3.21}$$

$$\frac{dP}{dt} = k_2C \tag{3.22}$$

To facilitate comparison of the unscaled (dimensioned) system with two different dimensionless versions of this problem discussed below, we include a switch parameter, called IFDIM. When this parameter has a value of 0, equations (3.20)—(3.22) are used. When IFDIM=1, then the following dimensionless equations, which are particularly appropriate in the "fast change" time domain, are used:

$$\frac{ds}{d\tau} = \frac{\tilde{E}_0 + \tilde{C}_0}{\tilde{S}_0 + 1}\left[-s + \frac{\tilde{S}_0}{\tilde{S}_0 + 1}\, cs + \frac{\eta(\eta+1)^{-1}}{\tilde{S}_0 + 1}\, c\right] \tag{3.23}$$

$$\frac{dc}{d\tau} = s - \frac{\tilde{S}_0}{\tilde{S}_0 + 1}\, cs - \frac{1}{\tilde{S}_0 + 1} c \tag{3.24}$$

$$\frac{dp_i}{d\tau} = c \tag{3.25}$$

τ stands for the dimensionless time:

$$\tau = (k_1 S_0 + k_{-1} + k_2)\ t \tag{3.26}$$

The dimensionless dependent variables are

$$s \equiv S/S_0, \qquad c \equiv C/\bar{C} \qquad p_i \equiv P/(k_2\bar{C}k^{-1}) \tag{3.27}$$

where

$$\bar{C} \equiv \frac{(E_0 + C_0)S_0}{K_m + S_0} \tag{3.28}$$

is the approximate "initial" complex concentration of *(MDP:4.22, p58)*, and

$$K_m = \frac{k_{-1} + k_2}{k_1} .$$

The four dimensionless parameters are

$$\tilde{C}_0 \equiv C_o/K_m, \quad \tilde{E}_0 \equiv E_0/K_m, \quad \tilde{S}_0 = S_0/K_m, \quad \eta = k_{-1}/k_2. \tag{3.29}$$

When $IFDIM = 2$, then the following dimensionless equations, which are particularly appropriate in the slow change, long time domain, are used:

$$\frac{\mathrm{d}s}{\mathrm{d}T} =$$

$$(\eta + 1)(\tilde{S}_0 + 1)\frac{\tilde{E}_0 + \tilde{C}_0}{\tilde{E}_0}\left[-s + \frac{\tilde{S}_0}{\tilde{S}_0 + 1}\,cs + \frac{\eta(\eta+1)^{-1}}{\tilde{S}_0 + 1}\,c\right] \tag{3.30}$$

$$\frac{\mathrm{d}c}{\mathrm{d}T} = \frac{(\eta + 1)(\tilde{S}_0 + 1)^2}{\tilde{E}_0}\left[s - \frac{\tilde{S}_0}{\tilde{S}_0 + 1}\,cs - \frac{1}{\tilde{S}_0 + 1}c\right] \tag{3.31}$$

$$\frac{\mathrm{d}p}{\mathrm{d}T} = \frac{\tilde{E}_0 + \tilde{C}_0}{\tilde{E}_0}c \tag{3.32}$$

Here the dimensionless time and product concentration are given by

$$T = \frac{k_2\tilde{E}_0}{\tilde{S}_0 + 1}\,t, \qquad p \equiv P/S_0 \ . \tag{3.33}$$

The names of the parameters differ in the three cases. Parameters for all three cases are included in the same F7≻[u] submenu[5] and in the table below. When you have IFDIM set equal to 1 or 2, the F7≻[i] option can compute the values of the parameters, $\tilde{C}_0$, $\tilde{E}_0$, $\tilde{S}_0$, and η,

[5] F7≻[u] denotes the [u] choice, made by striking the 'u' key, from the F7 menu.

in the dimensionless version of the equations, from the values of the dimensioned parameters in equations (3.20)—(3.22).

Component 1 is C or c, component 2 is S or s, and component 3 is P, p_i, or p. *Make sure you use the F7≻[i] option after every change of parameters for this problem, and do that* **after** *you have set the initial condition,* $S_0 = S(0)$*, from which* $\tilde{S}_0$ *is computed.*

You can toggle between the full third order version and the quasi–steady second order version of the enzyme–substrate system by alternating between *select* = 7 and *select* = 8 (see page 44)[6].

parameter in manual	computer name	ID number
E_0	e0	1
C_0	c0	2
k_1	kp1	3
k_{-1}	km1	4
k_2	k2	5
$\tilde{E}_0$	E0tild	6
$\tilde{S}_0$	S0tild	7
$\tilde{C}_0$	C0tild	15
η	eta	8
K_m	Km	10
IFDIM	IFDIM	9
	select	39

Associated files supplied		
.par	.pic	.key
enz_sub		

NOTES

[6]That is, use the [u] choice in the F7 menu to change the value of the parameter called *select* to either 7 or 8.

3.11 Enzyme–substrate problem, second order approximation

Select = 8 identifies the second order quasi–steady–state approximation of the enzyme substrate problem consisting of equations *(MDP:4.24)* and *(MDP:4.25, p59)*, in which C is given by the algebraic quasi-steady constraint *(MDP:4.23, p58)*.

BIOGRAPH plots only solutions to differential equations, not algebraic equations. Thus, in order to obtain plots of $C(t)$, we convert the algebraic equation *(MDP:4.23)*

$$C = \frac{E_0 + C_0}{K_m + S} S \quad , \tag{3.34}$$

to a differential equation by differentiating as follows:

$$\frac{dC}{dt} = \frac{d}{dt}\left(\frac{E_0 + C_0}{K_m + S} S\right) = \frac{K_m(E_0 + C_0)}{(K_m + S)^2} \frac{dS}{dt} \tag{3.35}$$

When equation *(MDP:4.24, p59)* is substituted for dS/dt, we obtain this system of three differential equations:

$$\frac{dS}{dt} = -\frac{k_2(E_0 + C_0)S}{K_m + S} \tag{3.36}$$

$$\frac{dP}{dt} = \frac{k_2(E_0 + C_0)S}{K_m + S} \tag{3.37}$$

$$\frac{dC}{dt} = -K_m k_2 \frac{(E_0 + C_0)^2}{(K_m + S)^3} S \tag{3.38}$$

Here the parameter K_m stands for:

$$K_m = \frac{k_{-1} + k_2}{k_1} \quad . \tag{3.39}$$

Equation (3.38) makes sense only if you enforce the initial condition:

$$C(0) = \frac{(E_0 + C_0)S(0)}{K_m + S(0)} \quad . \tag{3.40}$$

If, after specifying new parameter values using the F7≻[u] option, you pick the F7≻[i] option, that initial condition for $C(0)$ will be computed and stored automatically. If you forget to set this initial condition for $C(0)$, the program detects this and quickly adjusts $C(t)$ to satisfy equation *(MDP:4.23, p58)*.

Note that the parameters are identical to those in the previous (*select* = 7) problem. Thus, you can toggle between the full third order version and the quasi–steady second order version of the enzyme–substrate system by alternating between *select* = 7 and *select* = 8. When $IFDIM = 0$, equations (3.36)—(3.38) are used. We supply two dimensionless versions of equations (3.36)—(3.38). One set corresponds to equations (3.23)—(3.25) and is used when $IFDIM = 1$. Another set corresponds to equations (3.30)—(3.32)and is used when $IFDIM = 2$. We now derive these dimensionless versions.

When $IFDIM = 1$ or 2, we set the right side of equation (3.24) equal to 0 to obtain

$$c = \frac{1 + \tilde{S}_0}{1 + \tilde{S}_0 s}\, s \tag{3.41}$$

Differentiating, we obtain:

$$\frac{\mathrm{d}c}{\mathrm{d}\tau} = \frac{1 + \tilde{S}_0}{(1 + \tilde{S}_0 s)^2}\, \frac{\mathrm{d}s}{\mathrm{d}\tau} \tag{3.42}$$

τ can be replaced by T and equation (3.42) remains valid.

When $IFDIM = 1$, the following dimensionless equations, that are analogous to (3.23)—(3.25), are used:

$$\frac{\mathrm{d}s}{\mathrm{d}\tau} = -\frac{(\tilde{E}_0 + \tilde{C}_0)\, s}{(1 + \tilde{S}_0)(1 + \eta)(1 + \tilde{S}_0 s)} \tag{3.43}$$

$$\frac{\mathrm{d}p_i}{\mathrm{d}\tau} = \frac{1 + \tilde{S}_0}{1 + \tilde{S}_0 s}\, s \tag{3.44}$$

$$\frac{\mathrm{d}c}{\mathrm{d}\tau} = -\frac{(\tilde{E}_0 + \tilde{C}_0)\, s}{(1 + \tilde{S}_0 s)^3 (1 + \eta)} \tag{3.45}$$

Equation (3.45) is obtained by substituting equation (3.43) into equation (3.42).

When $IFDIM = 2$, the following dimensionless equations, that are analogous to (3.30)—(3.32), are used:

$$\frac{\mathrm{d}s}{\mathrm{d}T} = -\frac{(1+\tilde{S}_0)(\tilde{E}_0+\tilde{C}_0)\ s}{\tilde{E}_0(1+\tilde{S}_0 s)} \tag{3.46}$$

$$\frac{\mathrm{d}p}{\mathrm{d}T} = \frac{(\tilde{E}_0+\tilde{C}_0)}{\tilde{E}_0}\ \frac{1+\tilde{S}_0}{1+\tilde{S}_0 s}\ s \tag{3.47}$$

$$\frac{\mathrm{d}c}{\mathrm{d}T} = -\ \frac{(1+\tilde{S}_0)^2(\tilde{E}_0+\tilde{C}_0)\ s}{\tilde{E}_0(1+\tilde{S}_0 s)^3} \tag{3.48}$$

Equation (3.48) results from substitution of equation (3.46) into equation (3.42). The four dimensionless parameters are defined by

$$\tilde{C}_0 \equiv C_0/K_m, \quad \tilde{E}_0 \equiv E_0/K_m, \quad \tilde{S}_0 = S_0/K_m, \quad \eta = k_{-1}/k_2.$$

Equations (3.45) or (3.48) make sense only if you enforce the initial condition

$$c(0) = \frac{1+\tilde{S}_0}{1+\tilde{S}_0 s(0)}\ s(0) \tag{3.49}$$

If, after specifying new parameter values using the F7≻[u] option, you pick the F7≻[i] option, that initial condition for $c(0)$ will be computed and stored automatically. If you forget to set this initial condition for $c(0)$, the program detects this and quickly adjusts $c(t)$ to the appropriate value.

Component 1 is C or c, component 2 is S or s, and component 3 is P, p_i, or p.

parameter in manual	computer name	ID number
E_0	e0	1
C_0	c0	2
k_1	kp1	3
k_{-1}	km1	4
k_2	k2	5
$\tilde{E}_0$	E0tild	6
$\tilde{S}_0$	S0tild	7
$\tilde{C}_0$	C0tild	15
η	eta	8
K_m	Km	10
$IFDIM$	IFDIM	9
	select	39

Associated files supplied		
.par	.pic	.key
q_steady		

NOTES

3.12 Second order glycolytic oscillation problem

Select = 9 identifies the following second order glycolytic oscillation model:

$$\frac{d\alpha}{dt} = 1 - \sigma\Phi(\alpha, \beta) \tag{3.50}$$

$$\frac{d\beta}{dt} = q\sigma\Phi(\alpha, \beta) - k_t\beta \tag{3.51}$$

where

$$\Phi(\alpha, \beta) = \frac{\alpha(1+\alpha)^{n-1}(1+\beta)^n + L\theta\omega\alpha(1+\omega\alpha)^{n-1}}{L(1+\omega\alpha)^n + (1+\alpha)^n(1+\beta)^n} \tag{3.52}$$

This system of equations does not appear in MDP but is described in Exercises $\aleph 21_6$—$\aleph 27_6$ in chapter 7 beginning on page 125 in this manual[7]. Component 1 is β and component 2 is α.

parameter in manual	computer name	ID number
σ	sigma	1
q	q	2
k_t	kt	3
L	L	4
ω	w	5
θ	theta	6
n	nunits	7
	select	39

Associated files supplied		
.par	.pic	.key
glycosc	glycosc	glycosc

↝ Note that the parameter n represents the number of subunits involved in a reaction, *not* the order of the ode system. Only integer values of the parameter n make sense. Even though you can enter any floating point value for this parameter from the F7≻[u] menu option, the value used by the BIOGRAPH program in evaluating equations (3.50)—(3.52) is the greatest integer $\leq n$

[7]The notation $\aleph 22_6$ denotes exercise number 22 in this manual associated with chapter 6 in MDP.

3.13 Decroly–Goldbeter coupled feedback loop oscillator problem

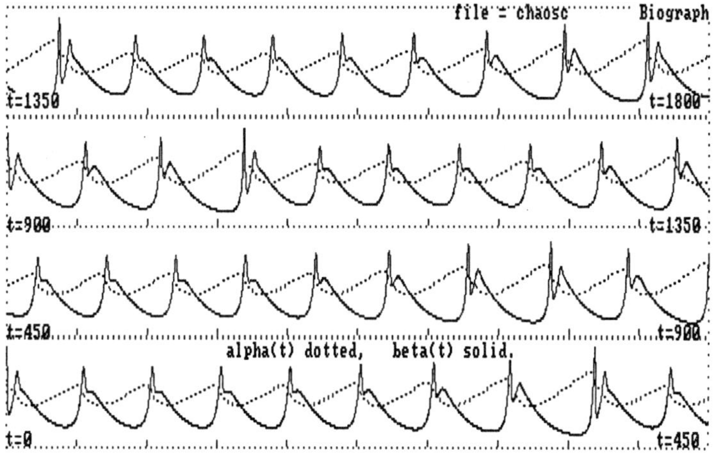

Figure 3.2: Chaotic oscillations of the Decroly–Goldbeter system. The parameter values can be found in the file called chaosc.par, the picture is stored in the file called chaosc.pic, and playback of the file called chaosc.key generates the results seen here.

$Select = 10$ identifies the following third order system:

$$\frac{d\alpha}{dt} = 1 - \sigma_1\phi(\alpha, \beta), \tag{3.53}$$

$$\frac{d\beta}{dt} = q_1\sigma_1\phi(\alpha, \beta) - \sigma_2\eta(\beta, \gamma) \tag{3.54}$$

$$\frac{d\gamma}{dt} = q_2\sigma_2\eta(\beta, \gamma) - k\gamma \tag{3.55}$$

where

$$\phi(\alpha, \beta) \equiv \frac{\alpha(1+\alpha)(1+\beta)^2}{L_1 + (1+\alpha)^2(1+\beta)^2} \tag{3.56}$$

$$\eta(\beta, \gamma) \equiv \frac{\beta(1+\beta d)(1+\gamma)^2}{L_2 + (1+\beta d)^2(1+\gamma)^2} \tag{3.57}$$

Exercises $\aleph 28_6$—$\aleph 34_6$ in chapter 7 of this manual concern this system[8]. This system does not appear in MDP. Component 1 is β, component 2 is α, and component 3 is γ.

parameter in manual	computer name	ID number
σ_1	sig1	1
σ_2	sig2	2
q_1	q1	3
q_2	q2	4
k	k	5
L_1	L1	6
L_2	L2	7
d	d	8
	select	39

Associated files supplied		
.par	.pic	.key
chaosc	chaosc	chaosc goldhart

↝ Note that playback of the file *chaosc.key* takes considerable computation time[9] because it involves integration from $t = 0$ to $t = 1800$, during which the tens of very complex wave forms seen in Figure 3.2 occur. When dealing with ODE systems that exhibit deterministic chaos behavior, as this one does, the truncation error bound parameter, *etol*, must be set to a very small value (10^{-8}, say). This is because even the slightest errors in tracing trajectories will be amplified eventually into large errors that give a false prediction about the *qualitative behavior* of a solution. Smaller values of *etol* postpone, but do not eliminate this defect. Small values of *etol*, however, cause lengthy computations.

NOTES

[8] The notation $\aleph 33_6$ denotes exercise number 33 in this manual associated with chapter 6 in MDP.

[9] It takes about nine minutes on a PC or PC-XT with the 8087 coprocessor chip. Without that chip, it would take one hour and forty minutes.

3.14 Second order calcium trigger, acto-myosin problem

Select = 11 identifies the second order system of equations *(MDP:9.7)* and *(MDP:9.8, p208)*, modified by naming the parameters α and β differently. α and β in MDP are replaced by (α'/β') and $1/\beta'$ below.

$$\frac{\mathrm{d}C}{\mathrm{d}t} = \gamma L - \nu C + \frac{\alpha' C^2}{1+\beta' C^2} \tag{3.58}$$

$$\frac{\mathrm{d}L}{\mathrm{d}t} = \frac{k(L,C)}{\mu(C)}\{L_0(C) - L\} \tag{3.59}$$

The following functions define $k(L,C)$, $\mu(C)$, and $L_0(C)$:

$$k(L,C) \equiv \frac{k_{max}}{1+k_2 L_0(C)^2 + k_3 L^2} \tag{3.60}$$

$$\mu(C) \equiv \mu_0 + \frac{\mu_{max} - \mu_0}{1+\left(\frac{C}{C_{melt}}\right)^2} \tag{3.61}$$

$$L_0(C) \equiv \epsilon + \frac{E}{1+(\frac{C}{C_{trig}})^2} \tag{3.62}$$

The cytogel viscosity, $\mu(C)$, varies from a maximum of μ_{max}, when $C = 0$, to a minimum of μ_0, as $C \longrightarrow \infty$. The elastic modulus k is an increasing function of C because $L_0(C)$ is a decreasing function of C. $L_0(C)$ falls from a maximum value of $\epsilon + E$ when $C = 0$, toward a minimum of ϵ as $C \longrightarrow \infty$. The equations are to be interpreted as normalized so that the longer steady–state length of the contractile filament is $L = 1$ when the calcium ion concentration, C, is at its lower steady state value. In MDP, the effects of increasing the calcium sensitivity parameter, α', are studied. When α' is very small, we want the system of differential equations to have a steady state solution with $L = L_0 = 1$. When $\alpha' = 0$, the C–nullcline has the formula. $L = \nu C/\gamma$, so that $L = 1$ is a steady state value only if $(C, L) = (\gamma/\nu, 1)$ is a steady–state point. $L = L_0(C)$ is the equation of the L–nullcline. We adjust the constant E so that the nullclines intersect at $L = L_0(\gamma/\nu) = 1$. This requires:

$$E = (1-\epsilon)\left\{1 + \left(\frac{\gamma}{\nu C_{trig}}\right)^2\right\} \tag{3.63}$$

You do not set the parameter E; the BIOGRAPH program automatically computes this value for E from the parameter values you actually supply. Component 1 is C and component 2 is L.

parameter in manual	computer name	ID number
α'	alphapr	1
β'	betapr	2
γ	gamma	3
ν	nu	4
C_{trig}	ctrig	5
C_{melt}	cmelt	6
ϵ	eps	7
μ_0	viscmin	8
μ_{max}	vscmax	9
k_{max}	kmax	10
k_2	k2	11
k_3	k3	12
	select	39

Associated files supplied		
.par	.pic	.key
catrig focal	bistab	bistab

NOTES

3.15 Generic third order system with quadratic right hand sides

Select = 12 identifies a generic third order system whose right hand sides are quadratic polynomials with a single cubic term[10]. This system does not appear in MDP but is required to solve some of the miscellaneous exercises in chapter 7 (see page 200). The file called *lorenz.par* contains parameter specifications that turn this system into the well known "Lorenz strange attractor" equations, which exhibit deterministic chaos. The general equations have this form:

$$\frac{\mathrm{d}x}{\mathrm{d}t} = a_0 + a_1 x + a_2 y + a_3 z + a_4 xy + a_5 xz + a_6 yz + a_7 x^2 + a_8 y^2 + a_9 z^2 + a_{10} x^3 \tag{3.64}$$

$$\frac{\mathrm{d}y}{\mathrm{d}t} = b_0 + b_1 x + b_2 y + b_3 z + b_4 xy + b_5 xz + b_6 yz + b_7 x^2 + b_8 y^2 + b_9 z^2 + b_{10} y^3 \tag{3.65}$$

$$\frac{\mathrm{d}z}{\mathrm{d}t} = c_0 + c_1 x + c_2 y + c_3 z + c_4 xy + c_5 xz + c_6 yz + c_7 x^2 + c_8 y^2 + c_9 z^2 + c_{10} z^3 \tag{3.66}$$

Component 1 is x, 2 is y, and 3 is z. The Lorenz equations are usually written, using three parameters, σ, r, and b, as follows:

[10] useful for treating the system of ODE's that results when exact travelling wave solutions to the Fitzhugh–Nagumo partial differential equations are sought

$$\frac{dx}{dt} = \sigma(y - x) \tag{3.67}$$

$$\frac{dy}{dt} = rx - y - xz \tag{3.68}$$

$$\frac{dz}{dt} = xy - bz \tag{3.69}$$

To turn equations (3.64)—(3.66) into the Lorenz equations, set all a_i, b_i, and c_i equal to zero except for:

$$a_1 = -\sigma, \qquad a_2 = \sigma, \qquad b_1 = r, \qquad c_3 = -b,$$

$$b_2 = -1, \qquad b_5 = -1, \qquad c_4 = 1.$$

One set of parameter values that yields chaotic behavior is $\sigma = 10.0$, $b = 2.66666667$, and $r = 28.0$. Play back the file called *lorenz.key* to see this chaotic behavior.

After playing back the *lorenz.key* file, listed below, try playing back the *owl.key* file which re-graphs the solution computed by *lorenz.key* in a two dimensional projection of the three dimensional phase portrait. The resulting picture is surprising.

parameter in manual	computer name	ID number
a_0	a0	1
a_1	a1	2
a_2	a2	3
a_3	a3	4
a_4	a4	5
a_5	a5	6
a_6	a6	7
a_7	a7	8
a_8	a8	9
a_9	a9	10
a_{10}	a10	32
b_0	b0	11
b_1	b1	12
b_2	b2	13
b_3	b3	14
b_4	b4	15
b_5	b5	16
b_6	b6	17
b_7	b7	18
b_8	b8	19
b_9	b9	20
b_{10}	b10	33
c_0	c0	21
c_1	c1	22
c_2	c2	23
c_3	c3	24
c_4	c4	25
c_5	c5	26
c_6	c6	27
c_7	c7	28
c_8	c8	29
c_9	c9	30
c_{10}	c10	34
	select	39

Associated files supplied		
.par	.pic	.key
lorenz		lorenz owl

3.16 Gierer–Meinhardt activator–inhibitor cell ring simulation problem

Select = 19 identifies a simulation of the second order activator–inhibitor reaction–diffusion problem consisting of the partial differential equations *(MDP:8.1a & 8.1b, p150)*

$$\frac{\partial a}{\partial t} = \rho_a(x)\left(c_a + c\frac{a^2}{h}\right) - \mu_a a + D_a\frac{\partial^2 a}{\partial x^2} \tag{3.70}$$

$$\frac{\partial h}{\partial t} = \rho_h(x)c_h a^2 - \mu_h h + D_h\frac{\partial^2 h}{\partial x^2} \tag{3.71}$$

$\rho_a(x)$ and $\rho_h(x)$ may vary with position, x, to account for a possible spatial pre–pattern in the morphogen source strengths.

The *select* = 6 case on page 40 deals with the ordinary differential equations that result when the diffusion terms are removed. Thus *select* = 6 can be used to study the properties of the morphogen biochemistry each individual cell would carry out in the absence of neighbors (*i.e.* when $D_h = D_a = 0$). Here we consider a ring of M cells (or intervals), with $2 \le M \le 8$. These cells are numbered $1, 2, ..., M$, arranged so that cell i has cells $(i-1)^*$ and $(i+1)^*$ as its counterclockwise and clockwise neighbors[11]. We seek a discrete approximation of equations (3.70)—(3.71) in the form of $2 \times M$ functions of time

$$h_1(t), a_1(t), \quad h_2(t), a_2(t) \ , \ . \ . \ . \ , \ h_M(t), a_M(t)$$

Here $(h_i(t), a_i(t))$ denotes the concentrations of inhibitor and activator in cell i. As in *(MDP:8.48, p161)*, we approximate the diffusion terms in equations by (3.70)—(3.71)

$$D_a\frac{\partial^2 a}{\partial x^2}(x_i) \quad \approx \quad \frac{D_a}{\delta^2}\left(a_{(i-1)^*} - 2a_i + a_{(i+1)^*}\right)$$

Here x_i is the coordinate of the ith cell's center and $\delta = \Delta x$ is the width of the "mesh" into which space is divided: For each i,

[11] For i between 2 and M, $(i-1)^*$ has the value $i-1$. For i between 1 and $M-1$, $(i+1)^*$ has the value $i+1$. To account for the ring topology, in which cell 1 is the clockwise neighbor of cell M, and in which cell M is the counterclockwise neighbor of cell 1, we assign the value M to $(1-1)^*$ and the value 1 to $(M+1)^*$.

$x_{i+1} - x_i = \delta$. Alternatively, if we take $\delta = 1$, we can regard our equations as describing activator–inhibitor reactions together with intercellular permeation of the chemicals between adjacent cells. See the last paragraph on p 161 of MDP. We obtain a system of $2 \times M$ ODE's to approximate the two partial differential equations (3.70)—(3.71), namely

$$\frac{\mathrm{d}a_i}{\mathrm{d}t} = \rho_{a_i}\left(c_a + c\frac{a_i^2}{h_i}\right) - \mu_a a_i + \frac{D_a}{\delta^2}\left(a_{(i-1)^*} - 2a_i + a_{(i+1)^*}\right) \tag{3.72}$$

$$\frac{\mathrm{d}h_i}{\mathrm{d}t} = \rho_{h_i} c_h a_i^2 - \mu_h h_i + \frac{D_h}{\delta^2}\left(h_{(i-1)^*} - 2h_i + h_{(i+1)^*}\right) \tag{3.73}$$

for $i = 1, ..., M$. The $2 \times M$ dependent variables are arranged in the ODE system solved by the BIOGRAPH program as follows. Component 1 is h_1, component 2 is a_1, component 3 is h_2, component 4 is a_2, . . . , component $2M - 1$ is h_M, and component $2M$ is a_M.

Two *.par* files are supplied. The one called *ring4.par* sets up a simulation with $M = 4$ cells in a ring, thus 8 ODE's. The one called *ring8.par* sets up a simulation with $M = 8$ cells, thus 16 ODE's. You can change M to any number between 2 and 8 by editing either *.par* file and changing the number on its first line to be twice M. Once you have read a *ring.par* file, *you cannot change M from within the BIOGRAPH program;* you must use a text editor program to change M. The F7≻[i] option will tell you how many cells are in the ring.

Note that the initial condition for each h_i must be a non-zero positive value because of the a_i^2/h_i terms in (3.72). If you mistakenly assign a zero initial condition to any h_i the program will replace it by 0.01 .

↝ To eliminate the need to type in identical values for all the ρ_{a_i} and ρ_{h_i} when simulating a spatially homogeneous domain, the following conventions are adopted. When $\rho_a \geq 0$, then for each $i = 1, 2, ..., M$, ρ_{a_i} is set equal to ρ_a inside the BIOGRAPH program, while the values specified in the F7≻[u] option for the ρ_{a_i} are ignored. When, however, $\rho_a < 0$, that negative value is ignored and the individual value specified in the F7≻[u] option for each ρ_{a_i} determines the activator strength in cell number i. Similarly, $\rho_h \geq 0$ means each ρ_{h_i} is set equal to ρ_h inside the BIOGRAPH program, and when $\rho_h < 0$, that negative value is ignored and the individual value specified in the F7≻[u] option for each ρ_{h_i} determines the inhibitor strength in cell number i.

⇝ The maximum number of sample points that can be stored for plotting during one integration process is 2000 if $M < 4$. When $M \geq 4$, the maximum number of sample points that can be stored is $16,535/(2M+1)$. The program enforces these limits automatically; you need not be careful about them. This note is provided only to explain why an integration might get cut short.

parameter in manual	computer name	ID number
c_a	ca	1
c	c	2
μ_a	mua	3
D_a	Da	4
c_h	ch	5
μ_h	muh	6
D_h	Dh	7
ρ_a	rhoa	8
ρ_h	rhoh	10
δ	deltax	11
ρ_{h_1}	rh1	12
ρ_{h_2}	rh2	13
ρ_{h_3}	rh3	14
ρ_{h_4}	rh4	15
ρ_{h_5}	rh5	16
ρ_{h_6}	rh6	17
ρ_{h_7}	rh7	18
ρ_{h_8}	rh8	19
ρ_{a_1}	ra1	20
ρ_{a_2}	ra2	21
ρ_{a_3}	ra3	22
ρ_{a_4}	ra4	23
ρ_{a_5}	ra5	24
ρ_{a_6}	ra6	25
ρ_{a_7}	ra7	26
ρ_{a_8}	ra8	27
	select	39

Associated files supplied		
.par	.pic	.key
ring4 ring8		

3.17 Cell ring diffusion simulation with general cubic reaction terms

Select = 20 identifies a simulation of the following system of reaction–diffusion equations:

$$\frac{\partial u}{\partial t} = \sum_{k=0}^{3}\sum_{j=0}^{3-k} a_{kj}u^k v^j + D_u\frac{\partial^2 u}{\partial x^2} \qquad (3.74)$$

$$\frac{\partial v}{\partial t} = \sum_{k=0}^{3}\sum_{j=0}^{3-k} b_{kj}u^k v^j + D_v\frac{\partial^2 v}{\partial x^2} \qquad (3.75)$$

These consist of the ordinary differential equations (3.16) and (3.17) in the *select* = 5 case on page 37 with diffusion terms added. Thus the *select* = 5 cases can be used to study the behavior of the spatially homogeneous system. The summation notation in (3.74) and (3.75) is a shorthand for

$$\sum_{k=0}^{3}\sum_{j=0}^{3-k} a_{kj}u^k v^j = a_{00} + a_{01}v + a_{02}v^2 + a_{03}v^3 + a_{10}u$$
$$+a_{11}uv + a_{12}uv^2 + a_{20}u^2 + a_{21}u^2v + a_{30}u^3$$

$$\sum_{k=0}^{3}\sum_{j=0}^{3-k} b_{kj}u^k v^j = b_{00} + b_{01}v + b_{02}v^2 + b_{03}v^3 + b_{10}u$$
$$+b_{11}uv + b_{12}uv^2 + b_{20}u^2 + b_{21}u^2v + b_{30}u^3$$

The partial differential equations (3.74) and (3.75) are converted to a system of ordinary differential equations corresponding to a ring of from 2 to 8 cells in exactly the same way as in the preceding *select* = 19 case[12]. $(u_i(t), v_i(t))$ denotes the concentrations of (u, v) in cell number i

[12]Refer to the preceding case for a more thorough discussion than is given here.

at time t. The finite difference approximations of equations (3.74) and (3.75) are

$$\frac{\mathrm{d}u_i}{\mathrm{d}t} = \sum_{k=0}^{3}\sum_{j=0}^{3-k} a_{kj}u_i^k v_i^j \quad + \quad \frac{D_u}{\delta^2}\left(u_{(i-1)^\star} - 2u_i + u_{(i+1)^\star}\right) \tag{3.76}$$

$$\frac{\mathrm{d}v_i}{\mathrm{d}t} = \sum_{k=0}^{3}\sum_{j=0}^{3-k} b_{kj}u_i^k v_i^j \quad + \quad \frac{D_v}{\delta^2}\left(v_{(i-1)^\star} - 2v_i + v_{(i+1)^\star}\right) \tag{3.77}$$

for $i = 1, ..., M$, where M is the number of cells in the ring.[13]. M cannot exceed 8 and must be at least 2. δ measures the width of each cell.

The $2 \times M$ dependent variables are arranged in the ODE system solved by the BIOGRAPH program as follows. Component 1 is u_1, component 2 is v_1, component 3 is u_2, component 4 is v_2, . . . , component $2M - 1$ is u_M, and component $2M$ is v_M.

One *.par* file called *cubring.par* is supplied that sets up a simulation with $M = 8$ cells in a ring, thus 16 ODE's. You can change M to any number between 2 and 8 by editing this file and changing the number on its first line to be twice M. Once you have read a *.par* file, *you cannot change M from within the BIOGRAPH program;* you must use a text editor program to change M. The F7≻[i] option will tell you how many cells are in the ring.

↝ The maximum number of sample points that can be stored for plotting during one integration process is 2000 if $M < 4$. When $M \geq 4$, the maximum number of sample points that can be stored is $16,535/(2M+1)$. The program enforces these limits automatically; you need not be careful about them. This note is provided only to explain why an integration might get cut short.

[13]For i between 2 and M, $(i-1)^\star$ has the value $i-1$. For i between 1 and $M-1$, $(i+1)^\star$ has the value $i+1$. To account for the ring topology, in which cell 1 is the clockwise neighbor of cell M, and in which cell M is the counterclockwise neighbor of cell 1, we assign the value M to $(1-1)^\star$ and the value 1 to $(M+1)^\star$.

parameter in manual	computer name	ID number
a_{00}	a00	1
a_{10}	a10	2
a_{01}	a01	3
a_{11}	a11	4
a_{20}	a20	5
a_{02}	a02	6
a_{21}	a21	7
a_{12}	a12	8
a_{30}	a30	9
a_{03}	a03	10
b_{00}	b00	11
b_{10}	b10	12
b_{01}	b01	13
b_{11}	b11	14
b_{20}	b20	15
b_{02}	b02	16
b_{21}	b21	17
b_{12}	b12	18
b_{30}	b30	19
b_{03}	b03	20
D_v	Dv	21
D_u	Du	22
δ	deltax	23
	select	39

Associated files supplied		
.par	.pic	.key
cubring		

NOTES

Chapter 4

Finite difference problems BIOGRAPH can solve

4.1 General information about difference equations

In the lists of particular finite difference problems presented in this chapter, we will describe finite difference equations in the general form in which they appear in MDP, namely in the form:

$$\begin{pmatrix} y_1[k+1] \\ y_2[k+1] \\ \cdot \\ \cdot \\ \cdot \\ y_n[k+1] \end{pmatrix} = \begin{pmatrix} g_1\left(y_1[k], y_2[k], \ldots, y_n[k], k\right) \\ g_2\left(y_1[k], y_2[k], \ldots, y_n[k], k\right) \\ \cdot \\ \cdot \\ \cdot \\ g_n\left(y_1[k], y_2[k], \ldots, y_n[k], k\right) \end{pmatrix} \qquad (4.1)$$

$y_i[k]$ denotes the value of the ith component at generation k.

Within the BIOGRAPH program, however, we use a different form for such equations. This detail, described in the next four paragraphs can be ignored on a first reading.

↝ While the qualitative behavior of systems of finite difference equations can be dramatically different from that of differential equation systems, the same kind of mathematical operations are involved to solve and graph both kinds of systems. Internally, the BIOGRAPH program treats the two kinds of systems almost identically. To facilitate similar treatment all systems of difference equations are written inside the program in the following form:

$$\begin{pmatrix} y_1[k+1] \\ y_2[k+1] \\ \cdot \\ \cdot \\ \cdot \\ y_n[k+1] \end{pmatrix} = \begin{pmatrix} y_1[k] + f_1\,(y_1[k], y_2[k], ..., y_n[k], k) \\ y_2[k] + f_2\,(y_1[k], y_2[k], ..., y_n[k], k) \\ \cdot \\ \cdot \\ \cdot \\ y_n[k] + f_n\,(y_1[k], y_2[k], ..., y_n[k], k) \end{pmatrix} \tag{4.2}$$

so that $\vec{f}\,(\vec{y}\,[k], k)$ denotes a generational rate of change.

The above form for writing difference equation systems emphasizes the similarity to differential equations:

- Steady state solutions are $\vec{y}\,[k] = \vec{y}^{\,0}$ for all k, where $\vec{y}^{\,0}$ is any solution of $\vec{f}\,(\vec{y}^{\,0}, k) \equiv 0$. Thus, the procedure for finding steady state points, initiated by the F5≻[c] option, gives valid information.
- For a system of two first order difference equations, in which the right hand sides have no explicit dependence on k, a phase portrait analog makes sense. Figure 4.3 on page 73 shows an example. In particular, the nullcline location process initiated by the F5≻[n] menu remains useful for difference equations.

The form (4.2) is included to make clear which functions the nullcline location and critical point finder algorithms use.

One distinct incompatibility between the two kinds of systems is that, while both forward and backward integration of differential equations are valid, ***only forward iteration of finite difference equations is possible***. The program will not let you iterate backwards.

When initial conditions, $\vec{y}\,[0]$, are prescribed, simple iteration calculates all future generations, $\vec{y}\,[1], \vec{y}\,[2], ..., \vec{y}\,[k], ...$, and these dependent variables can be plotted as functions of generation number, k. That standard plotting style suffices to demonstrate simple dynamics of finite difference equations, when, for example, the solutions tend to stable periodic orbits with small period. Play back the file called *lesson1.key* to see a simple example.

When, however, these deterministic models exhibit chaotic (almost stochastic) behavior, as discussed on pps 25 *et. seq.* of MDP, this standard plotting format no longer suffices. Instead, one wants to compute thousands of generations and exhibit each generation as a single isolated point. To this end, the F3 menu, which schedules variables for plotting and determines line styles, offers via the [p] option, the choice of drawing each solution datum as a single isolated point, eliminating the usual connecting lines. These points should be plotted on the analog of a phase portrait for a pair of autonomous difference equations. See Figure 4.3 on page 73 for an example, generated by the playback file *insect.key*.

Superimposed epochs, a stochastic histogram

This section is somewhat complicated. It explains how to generate the kind of plots shown in figures 4.1 and 4.2. Every difference equation system the BIOGRAPH program can solve has three "extra" parameters associated with it (called T_{ramp}, E_{left}, and E_{right}) and one "extra" dependent variable component (called "the ramp function"), that are used *only* to construct such plots. If you are not interested in such plots, skip to the next section. When specifying parameter values using the F7≻[u] option just ignore those three parameters, and make no use of the y component called the "ramp function".

To understand a revealing alternative way to graph solutions of difference equations, consider a variable, $y[k]$, that varies chaotically as the "generations" $k = 1, 2, 3, ...$ pass. Even though the variation of $y[k]$ *vs.* k. seems random, often there are hidden patterns. To see these it turns out to be helpful to choose an "epoch" of T generations in length, and to superimpose plots of several epochs on a narrow vertical strip of the plot. Several vertical strips, each corresponding to different parameter values, can be displayed side by side on the same graph. We have chosen $T = 200$ as a suitable standard epoch length, but you can change T to whatever value you please.

Figure 4.1 shows an example with six vertical strips for six different values of the parameter r in the simple population dynamics equation *(MDP:2.7, p17)*[1]. This problem is further discussed as a special case of the *select* = 21 problem on page 68. Each epoch comprises two hundred generations, and each vertical strip shows the superposition

[1]This refers to equation number 7 in chapter 2 of MDP on page 17.

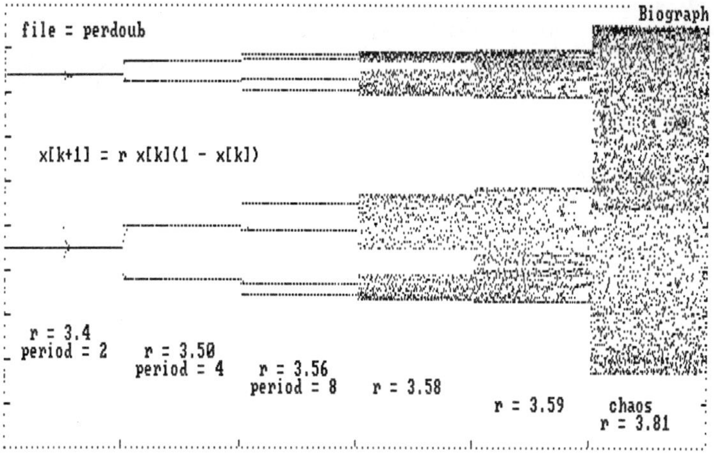

Figure 4.1: An example of superimposed epochs, illustrating the "period doubling route to chaos". The vertical panel widths are not proportional to the increments in the parameter r. This picture is generated by the file called perdoub.key and is stored in the file called perdoub.pic

of many such epochs. Note how clearly the periodic solutions stand out[2]. Note also that a certain "order in randomness" is demonstrated before the fully chaotic state is attained. This order, in particular, would be almost impossible to discern in a conventional graph of $y[k]$ as a function of k.

In Figure 4.2 we show how to generate the often used "pitchfork" bifurcation diagram for the same difference equation system. Here, one hundred epochs are graphed, each occupying a very narrow vertical panel. Figure 4.2 is a "high resolution" version of the simpler Figure 4.1.

↝ For those interested, we now explain the mathematical details of how the epoch superposition is done. Given an nth order system, one simply adds an additional component, y_{n+1}, having the form of a "ramp function", that repeats itself every T generations. The epoch length, e.g. $T = 200$, has no special significance. In each finite difference problem, we have introduced a ramp function and three new parameters, called E_{left} and E_{right} (to control the ramp amplitude), and T_{ramp} to specify its period, T. This ramp function rises linearly from a minimum value of E_{left} to a

[2] A solution of period N graphs as N sharp horizontal traces.

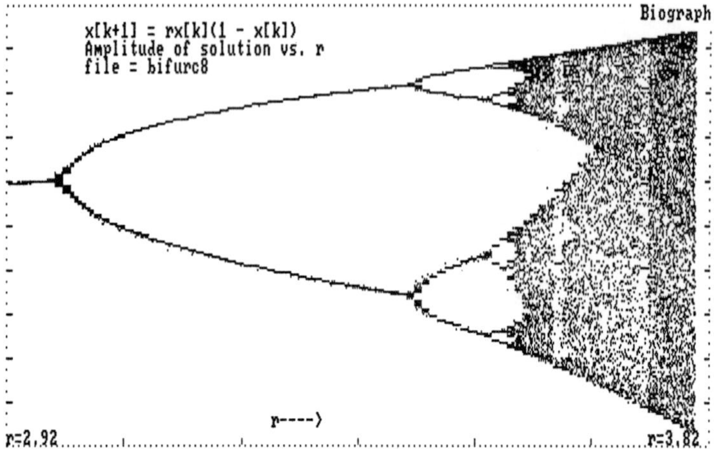

Figure 4.2: A "pitchfork" bifurcation diagram for equation (MDP:2.7 p17). The playback file called bifurc8.key produces this figure, which is stored in the file called bifurc8.pic.

maximum of E_{right} over its period of $T = T_{ramp}$ generations, then drops abruptly back to its minimum value, E_{left} to repeat.

To obtain a superimposed epoch graph of the ith component of the solution, y_i, proceed as follows:

1. Use the F3≻[H] option to specify y_{n+1} as the horizontal axis variable.
2. Use the F3≻[h] option to set the scale for the horizontal axis.
3. Use the F3≻[p] option to determine a fixed scale for the dependent variable(s) you want to plot on the vertical axis.
4. Pick a particular set of parameter values using the F7≻[u] option. In particular:
   - decide where on the horizontal scale you wish to locate the left and right edges of the vertical strip in which to superimpose several epochs of generations.
   - Using the F7≻[u] option, set the parameters E_{left} and E_{right} equal to the horizontal scale values that locate those left and right edges of the vertical strip. Set the parameter T_{ramp} to a reasonable epoch length, such as 200 generations. Beware that, for a given number of pixels between E_{left} and E_{right}, certain values of the ramp period will cause vertical banding in the plot.

The cure for this is to change T_{ramp}.

5. Use the F5≻[t] option to compute and plot $y_i[k]$ on the vertical axis as a function of $y_{n+1}[k]$, on the horizontal axis[3], for $k = 0, 1, 2, ..., \eta T$ generations, where ηT represents many, or at least several, epochs. If, for example, you compute $\eta T = 2000$ generations, this will generate 10 epochs of length $T_{ramp} = 200$ Because of step 1, these epochs will be superimposed, with the data for generations 1, 201, 401, 601, 801, etc. all occupying the same horizontal position on the plot specified by E_{left}. The data for generations 200, 400, 600, 800, 1000 will occupy the same horizontal position specified by E_{right}.

Repeat steps 4 and 5 for different sets of parameter values. Play back the file called *perdoub.key* to see the concrete example shown in Figure 4.1.

When you use the steady state point location capability of the F5≻[c] menu option, the $(n+1)^{st}$ dependent variable is "supressed" so that only steady state solutions of the "real" n^{th} order system are sought.[4]

4.2 Specific difference equation problems and their cases

The remainder of this chapter consists of a catalog of all the finite difference equations the BIOGRAPH program can solve. Recall our convention that equation numbers in this book are written in the style (4.7), while we use the following style to cite equations in MDP. *(MDP:4.3, p52)* refers to equation 3 in chapter 4 on page 52 of MDP. Sometimes, we omit the page references.

↝ Remember: Unless you want to employ the superimposed epochs feature (p64?) ignore the parameters $Tramp$, $Eleft$, and $Eright$. Also ignore the extra dependent variable, the ramp function. If you are asked to provide information about the ramp function, such as its initial value or its range, type *any* number .

[3] That is, on the segment of the horizontal axis specified by the parameters E_{left} and E_{right}

[4] It makes no sense to find a steady state point for a system that includes the ramp function, because it has none.

4.3 First order cubic polynomial difference equation

Select = 21 identifies a first order difference equation with a cubic polynomial right hand side:

$$y[k+1] = ay[k] + by[k]^2 + cy[k]^3 \tag{4.3}$$

For appropriate choices of the parameters a, b, and c, this equation subsumes the following problems in MDP:

- Equation *(MDP:2.7, p17)*, when $a = r$, $b = -r$, and $c = 0$:

  $$y[k+1] = ry[k](1 - y[k]) \tag{4.4}$$

  The files *lesson1.key*, perdoub.*, and bifurc8.* specify this equation. Play back of *perdoub.key* will produce Figure 4.1. The file called *bifurc8.key* causes a lengthy calculation and produces the standard "pitchfork" bifurcation diagram (see Figure 4.2) associated with this simple difference equation.

- Exercise 3 on page 32 of MDP, when $a = \frac{1}{4}$, $b = 1$, and $c = -\frac{1}{4}$:

  $$y[k+1] = y[k](1 + 4y[k] - y[k]^2)/4 \tag{4.5}$$

- Exercise 4(d) on page 32 in MDP, when, with a slight change of notation, $a = 1 - c$, and $b = 0$:

  $$y[k+1] = (1 - c)y[k] + cy[k]^3 \tag{4.6}$$

Component 1 is y and component 2 is the ramp function.

parameter in manual	computer name	ID number
a	a	1
b	b	2
c	c	3
T_{ramp}	tramp	36
E_{left}	eleft	37
E_{right}	eright	38
	select	39

Associated files supplied		
.par	.pic	.key
perdoub bifurc8	perdoub bifurc8	perdoub bifurc8 lesson1

4.4 First order exponential difference equation

Select = 22 identifies the following first order difference equation from exercise 4, page 32, of MDP:

$$y[k+1] = y[k]e^{s(1-y[k])} \tag{4.7}$$

Component 1 is y and component 2 is the ramp function.

parameter in manual	computer name	ID number
s	s	1
T_{ramp}	tramp	36
E_{left}	eleft	37
E_{right}	eright	38
	select	39

Associated files supplied		
.par	.pic	.key
dif22		

NOTES

4.5 Delayed difference equation: a second order system

Select = 23 identifies the delayed first order difference equation in exercise 5, page 32 in MDP:

$$x[k+1] = rx[k]\{1 - x[k-1]\} \tag{4.8}$$

This is equivalent to the following second order standard difference equation system:

$$y_1[k+1] = ry_1[k]\{1 - y_2[k]\} \tag{4.9}$$

$$y_2[k+1] = y_1[k] \tag{4.10}$$

Component 1 is $x[k]$, component 2 is $x[k-1]$, and component 3 is the ramp function. See the *select* = 28 problem for a more complicated delayed difference equation.

parameter in manual	computer name	ID number
r	r	1
T_{ramp}	tramp	36
E_{left}	eleft	37
E_{right}	eright	38
	select	39

Associated files supplied		
.par	.pic	.key
dif23		

NOTES

4.6 Second order host–parasitoid difference equation system

Select = 24 identifies a second order host–parasitoid model that is the subject of Exercises $\aleph 7_2$—$\aleph 10_2$.

$$X[n+1] = kX[n]e^{-a|Y[n]|^{1-m}} \tag{4.11}$$

$$Y[n+1] = kX[n]\left(1 - e^{-a|Y[n]|^{1-m}}\right) \tag{4.12}$$

Note that n counts the generation number and does *not* denote the order of the difference equation system. k is a parameter. Component 1 is $X[n]$, component 2 is $Y[n]$, and component 3 is the ramp function. $|y[n]|$ is the absolute value of $y[k]$[5]. In writing the equations on the screen, we use the common computer notation wherein * denotes multiplication and ** denotes exponentiation.

parameter in manual	computer name	ID number
k	kappa	1
a	a	2
m	m	3
T_{ramp}	tramp	36
E_{left}	eleft	37
E_{right}	eright	38
	select	39

Associated files supplied		
.par	.pic	.key
bupalus		

[5]The computer software will crash if asked to raise negative quantities to non-integral powers. In case $y[n] < 0.0$ is specified in initial conditions, $|y[n]|$ is used in the above equations.

4.7 First order population genetics difference equation

Select = 25 identifies the following first order population genetics model, which is equation *(MDP:3.6, p38)*, modified to include the effects of mutation:

$$p[m+1] = \frac{p[m](1-Kp[m])(1-\mu)+\nu(1-p[m])(1-(1-p[m])L)}{1-K(p[m])^2-L(1-p[m])^2} \tag{4.13}$$

μ measures the $A \to a$ mutation rate and ν measures the $a \to A$ mutation rate. When both μ and ν are set to zero, equation *(MDP:3.6, p38)* is recovered. Exercises $\aleph 4_3$—$\aleph 10_3$, starting on page 137 in chapter 7 deal with the derivation of, and simulations with, these mutation terms.

Note the use of $[m]$ to denote generation number to avoid confusion with the parameter K in equation (4.13). Component 1 is p and component 2 is the ramp function.

parameter in manual	computer name	ID number
K	K	1
L	L	2
μ	mu	3
ν	nu	4
T_{ramp}	tramp	36
E_{left}	eleft	37
E_{right}	eright	38
	select	39

Associated files supplied		
.par	.pic	.key
popgen		popgen

NOTES

4.8 Generic second order difference system with cubic right hand sides

Select = 26 identifies a general second order difference system with cubic polynomial right hand sides. This equation does not appear in MDP. It subsumes as special cases many popular population dynamics models. The equations have this form:

$$x[k+1] = \sum_{i=0}^{3} \sum_{j=0}^{3-i} a_{ij} x[k]^i y[k]^j \tag{4.14}$$

$$y[k+1] = \sum_{i=0}^{3} \sum_{j=0}^{3-i} b_{ij} x[k]^i y[k]^j \tag{4.15}$$

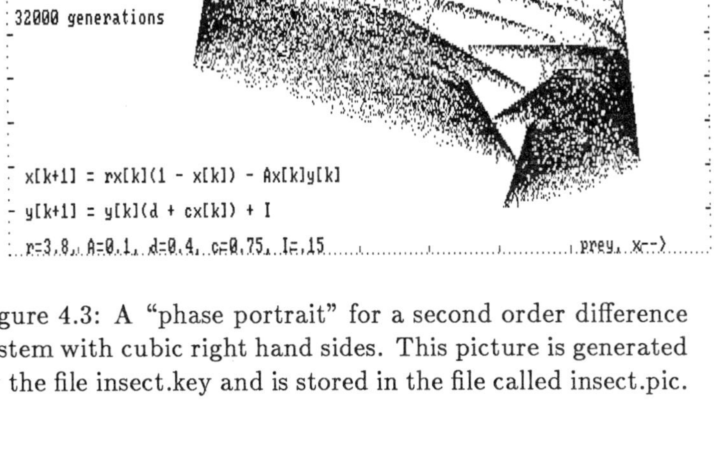

Figure 4.3: A "phase portrait" for a second order difference system with cubic right hand sides. This picture is generated by the file insect.key and is stored in the file called insect.pic.

The summation notation in equations (4.14) and (4.15) is shorthand for:

$$x[k+1] = a_{00} + a_{01}y + a_{02}y^2 + a_{03}y^3 + a_{10}x + a_{11}xy$$

$$+a_{12}xy^2 + a_{20}x^2 + a_{21}x^2y + a_{30}x^3$$

$$y[k+1] = b_{00} + b_{01}y + b_{02}y^2 + b_{03}y^3 + b_{10}x + b_{11}xy$$

$$+b_{12}xy^2 + b_{20}x^2 + b_{21}x^2y + b_{30}x^3$$

On the right hand sides of the equations above, x stands for x[k] and y stands for y[k]. Component 1 is x, component 2 is y, and component 3 is the ramp function.

See Figure 4.3 for an example of how to use this generic system. Use the player piano feature to play back the file called *insect.key* to see how to create that figure. The equations used in Figure 4.3 are

$$x[k+1] = rx[k](1 - x[k]) - Ax[k]y[k] \tag{4.16}$$

$$y[k+1] = y[k](d + cx[k]) + I \tag{4.17}$$

These model the population dynamics of a prey species, $x[k]$, and a predator, $y[k]$. In the absence of predators, the prey reproduce logistically. Each predator kills a number of prey proportional to the prey population each generation. I is a constant immigration rate for predators. The parameter d is adjusted to be less than 1.0 so that, without prey and without predator immigration, the predator population would go extinct. The values of the parameters used to generate this figure can be found in the file *insect.par*. To specialize the general equations (4.14) and (4.15) to those in the figure, set all a_{ij} and b_{ij} equal to zero, except:

$$a_{10} = r, \quad a_{20} = -r, \quad a_{11} = -A,$$
$$b_{00} = I, \quad b_{01} = d, \quad b_{02} = c \ .$$

parameter in manual	computer name	ID number
a_{00}	a00	1
a_{10}	a10	2
a_{01}	a01	3
a_{11}	a11	4
a_{20}	a20	5
a_{02}	a02	6
a_{21}	a21	7
a_{12}	a12	8
a_{30}	a30	9
a_{03}	a03	10
b_{00}	b00	11
b_{10}	b10	12
b_{01}	b01	13
b_{11}	b11	14
b_{20}	b20	15
b_{02}	b02	16
b_{21}	b21	17
b_{12}	b12	18
b_{30}	b30	19
b_{03}	b03	20
T_{ramp}	tramp	36
E_{left}	eleft	37
E_{right}	eright	38
	select	39

Associated files supplied		
.par	.pic	.key
insect	insect	insect

NOTES

4.9 First order rational function difference equation

Select = 27 identifies the following first order difference equation whose right hand side is the quotient of cubic polynomials:

$$x[k+1] = \frac{a_0 + a_1 x[k] + a_2(x[k])^2 + a_3(x[k])^3}{b_0 + b_1 x[k] + b_2(x[k])^2 + b_3(x[k])^3} \tag{4.18}$$

This equation does not appear in MDP.

parameter in manual	computer name	ID number
a_0	a0	1
a_1	a1	2
a_2	a2	3
a_3	a3	4
b_0	b0	5
b_1	b1	6
b_2	b2	7
b_3	b3	8
T_{ramp}	tramp	36
E_{left}	eleft	37
E_{right}	eright	38
	select	39

Associated files supplied		
.par	.pic	.key
ratdif		

NOTES

4.10 Second order exponential difference equation

Select = 28 identifies the following second order difference equation:

$$x[k+1] = a_0 x[k] + a_1(x[k-1])^2 + a_2 x[k] e^{(a_3 + a_4 x[k-1])} \tag{4.19}$$

This is equivalent to the following pair of first order difference equations.

$$y_1[k+1] = a_0 y_1[k] + a_1(y_2[k])^2 + a_2 y_1[k] e^{(a_3 + a_4 y_2[k])} \tag{4.20}$$

$$y_2[k+1] = y_1[k]. \tag{4.21}$$

Component 1 is $x[k]$ and component 2 is $x[k-1]$. Exercise number 6 on page 33 of MDP results when you set $a_0 = a_1 = 0$, $a_2 = r$, $a_3 = 1$, and $a_4 = -1$.

parameter in manual	computer name	ID number
a_0	a0	1
a_1	a1	2
a_2	a2	3
a_3	a3	4
a_4	a4	5
T_{ramp}	tramp	36
E_{left}	eleft	37
E_{right}	eright	38
	select	39

Associated files supplied		
.par	.pic	.key
expdelay		

NOTES

4.11 Generic third order system with quadratic right hand sides

Select = 29 identifies a generic third order system with quadratic polynomial right hand sides. This system does not appear in MDP. It is included here mainly for research use, for example to examine effects of delay. See Exercise $\aleph 3(g)_2$. The equations have this form:

$$x[k+1] = a_0 + a_1 x + a_2 y + a_3 z + a_4 xy + a_5 xz + a_6 yz + a_7 x^2 + a_8 y^2 + a_9 z^2 + a_{10} x^3 \quad (4.22)$$

$$y[k+1] = b_0 + b_1 x + b_2 y + b_3 z + b_4 xy + b_5 xz + b_6 yz + b_7 x^2 + b_8 y^2 + b_9 z^2 + b_{10} y^3 \quad (4.23)$$

$$z[k+1] = c_0 + c_1 x + c_2 y + c_3 z + c_4 xy + c_5 xz + c_6 yz + c_7 x^2 + c_8 y^2 + c_9 z^2 + c_{10} z^3 \quad (4.24)$$

On the right hand sides of equations (4.22)—(4.24) x, y, and z denote $x[k]$, $y[k]$, and $z[k]$, respectively. Component 1 is x, 2 is y, and 3 is z.

parameter in manual	computer name	ID number
a_0	a0	1
a_1	a1	2
a_2	a2	3
a_3	a3	4
a_4	a4	5
a_5	a5	6
a_6	a6	7
a_7	a7	8
a_8	a8	9
a_9	a9	10
a_{10}	a10	32
b_0	b0	11
b_1	b1	12
b_2	b2	13
b_3	b3	14
b_4	b4	15
b_5	b5	16
b_6	b6	17
b_7	b7	18
b_8	b8	19
b_9	b9	20
b_{10}	b10	33
c_0	c0	21
c_1	c1	22
c_2	c2	23
c_3	c3	24
c_4	c4	25
c_5	c5	26
c_6	c6	27
c_7	c7	28
c_8	c8	29
c_9	c9	30
c_{10}	c10	34
T_{ramp}	tramp	36
E_{left}	eleft	37
E_{right}	eright	38
	select	39

Associated files supplied		
.par	.pic	.key
diffquad		

Chapter 5

Actions offered by the menus

The program's major menus are listed below exactly as they appear on the screen. These are printed here in "verbose form". Each menu has a terse version with short names for the same choices. When you have become an "expert" (after using the program for a few hours), you can speed up the menu display operations by selecting these terse menus. The '!' key toggles between terse and verbose menu modes. Try to learn to use the terse form as soon as you can because this greatly quickens the response of the program.

The actions that will result from most menu choices are described adequately by the menu text itself. When appropriate, detailed descriptions of some choices in each menu follow the facsimile display of the menu.

The next section summarizes the functions invoked by each of the PC's ten function keys. Following that overview, a separate section is devoted to the details of each important submenu.

5.1 The main menu

Here is the main menu, each of whose options is invoked by hitting one of the keyboard's ten function keys, which are labeled F1, F2, ..., F10. The function key used to pick each option is displayed in the left margin.

F1 = create/clarify/embellish/save/retrieve a picture
F2 = erase the screen
F3 = define plotting style <<*MAIN MENU*>>
F4 = new initial conditions
F5 = integrate/draw solution curves/steady–state points/cursor
F6 = scan data files; print parameter/performance/case data
F7 = identify/get/change/save the problem-case
F8 = start/stop session log, playback, or keystroke saving
F9 = EXIT PROGRAM
F10 = display this menu

Choices F1, F3, F4, F5, F6, F7, F8 call secondary menus that perform concrete actions needed to carry out the major operations. Some of the choices in the secondary menus invoke tertiary menus, and so on. Moving to "deeper" menu levels gives you higher magnification views of details involved in implementing a task.

F10 is special. Selecting it when viewing the main menu just redisplays the main menu. Sometimes, however, nothing appears on the screen except the graph you are constructing, uncontaminated by any menu text or other prompts from the program. This is essential for printing out plot hardcopies and for an unobstructed view of the fruits of your labor. When the screen is "pure" in this way, or whenever one of the secondary menus is displayed, hitting F10 will recall the main menu shown above. This is true even though no screen message suggests it. The same is true of the F9 (quit the program) option in the main menu; you can quit by hitting the F9 key whenever you see any menu.[1]

Details of the main menu

F1 = create/clarify/embellish/save/retrieve a picture

See page 87 for a detailed description of the F1 submenu's options. Pick this to redraw the most recently computed solution curve(s), to save finished pictures in a disk file, or to recall previously stored pictures.

[1]You will be asked if you are sure you want to quit, so you have one chance to change your mind before commanding this irrevocable action.

One F1 pick lets you see the names of all files on any active disk drive. From the F1 menu, you can embellish a picture on the screen with text captions.

F2 = erase the screen

This erases the screen after asking you to confirm that irrevocable action[2]. To erase the screen you must get back to the main menu, hit the F2 key, and, finally, confirm your intention to erase. This process is intentionally awkward to avoid accidental erasure of a picture in which you have invested substantial effort. Get in the habit of using the F1≻[s] option[3] to save pictures before erasing them.

F3 = define plotting style

From the F3 menu you specify the format of plots drawn on the screen. See page 90 for a detailed description of the F3 submenu's options. You can use 1, 2, 3, or 4 "windows". Each window is a horizontal panel extending the width of the screen, with an optional border and optional horizontal and/or vertical tic marks on the axes. You specify the variable measured along the horizontal axis (the same one for all windows), and the minimum and maximum values on that axis. This is usually the independent variable, but can be one of the dependent variables (as is always the case in phase portraits).

You can plot any dependent variable component in any window using any dot-dash line style, and each dependent variable can be individually scaled by limits you prescribe. Alternatively the dependent variable can be automatically scaled or "normalized", meaning that the maximum is put at the top of the window and the minimum is put at the bottom. The F6≻[e] option tells you what these extremal values are.

When a solution is computed numerically (from the F5 menu) the data is stored before it is plotted so that the same data can be replotted in different formats without recomputation. Thus, having computed a solution, you can redesign the way you graph it using the F3 menu, then

[2] Each time the F2 key is hit, the program displays on the screen the (hexadecimal) addresses between which it is loaded in your computer's memory.

[3] "F1≻[s]" refers to the [s] choice made from the F1 submenu.

replot it from the F1 menu, and iterate until you see what you want to see. You control the number of sampled values stored during the integration process from the F3 menu; few sampled values give faster performance; many yield smoother plots. You can specify either that samples be saved at equal time intervals or that samples be taken at equally spaced increments along the solution trajectory.

F4 = new initial conditions

From the F4 menu you can change or inspect the initial conditions for each new solution you compute. If you are solving a second order system and constructing a phase portrait, you can invoke a graphics cursor from this menu to locate the initial point on a trajectory. See page 94 for a detailed description of the F4 submenu's options.

F5 = integrate/draw solution curves/steady–state points/..

This is the menu from which the mathematically important things are done. From the F5 menu you initiate the integration process, and possibly draw out the solution according to the specifications prescribed from the F3 menu. You can determine the initial and final times for which the solution is computed. The same menu serves analogous purposes for finite difference equations. See page 95 for a detailed description of the F5 submenu's options.

One pick from the F5 menu finds steady–state points and, for second order systems, classifies them. If you are making a phase plane portrait of a second order system, the F5 menu provides several special options: drawing arrowheads on trajectories; drawing grids of flow arrows; picking new initial conditions with a graphics cursor; finding and drawing nullclines.

F6 = scan data files; print parameter/performance-/case data

From the F6 menu you can get printed displays on the screen and, optionally, on a printer, of all current conditions. See page 100 for a detailed description of the F6 submenu's options. These include values of all relevant parameters, current plot format specification, and initial conditions. If you have computed a solution using the F5 menu, you can find out extrema attained by the independent variable and all dependent variable components. These are useful for specifying the plotting format from the F3 menu. You can list out entire problem/case **.par* files, which contain all the information to recompute and redraw the picture you have just constructed.

One purpose of this menu's options is to generate a printer listing of any desired information to keep with printer hardcopy of a successful picture you have produced.

F7 = identify/get/change/save the problem/case

See page 102 for a detailed description of the F7 submenu's options. Use this menu to read in previously stored problem/case files from the disk(s), or to write out the present problem/case file for future recall. You use this interactive program to explore the behavior of differential or difference equation systems. By the time you find interesting parameter values and the proper plotting format, etc., you may have invested substantial effort. The F7 menu provides an easy way to archive all this information in a single disk file so you can re–create your results without rediscovering them. You should use this feature copiously; it's easy and uses little disk space. A typical parameter file is only 2000 bytes long, so that hundreds of them will fit on one floppy diskette.

From the F7 menu you can view the names of the files now on your disk(s) without terminating the BIOGRAPH program. One of the F7 menu choices exhibits, then lets you change, the named parameters appearing in the right hand side of the ODE system. Another choice lets you view and change the parameters that control the Adams-Moulton variable order multi-step predictor-corrector algorithm that integrates the differential equations numerically.

From the F7 menu, you can invoke a screen message telling you what problem is presently active[4]. This tells you what pages in the text book describe the problem, and tells you which components of the solution have which names in MDP and in this manual. Sometimes it

[4]as determined by the last problem/case file you read in using the F7 menu

checks certain constraints among parameters appearing in the text or this manual and tells you if any is violated. For example, certain parameter values and initial conditions *must* be positive and the program enforces these constraints.

F8 = start/stop session log, playback, or keystroke saving

From this menu you initiate or terminate disk file recording of your interaction with the program. See page 113 for a detailed description of the F8 submenu's options. You can record all input and screen output in "session log" files. You can save just the input stream of keys you type in player piano **.key* files. You can play back previously recorded **.key* files. You should create for later replay your own library of **.key* files to store frequently used sequences of commands. *That is, in computer lingo, the F8 menu provides a command macro facility.*

F9 = EXIT PROGRAM

This provides a graceful exit from the BIOGRAPH program that restores the previous conditions of the operating system. The most important thing it does is replace the high resolution graphics display with the ascii text screen. Prior to quitting you should save your current problem/case file, using menu F7 if you think you might want to recall that case in a future session. Before taking the irrevocable action of quitting, the program asks you to confirm your intention.

Note that from any of the sub-menus you can hit the F9 key and terminate this program, even though this ever–present option is displayed only in the main menu.

> ↝ Any currently open session log or keystroke saving file will be closed before terminating the program if you use the 'F9' key exit procedure. The 'Ctrl' 'c' or 'Ctrl' 'Break' keys both abort the BIOGRAPH program gracelessly, leaving the graphics display active; they fail to save any session logging or **.key* file that happens to be open. If you use this mode to exit and wish to restore the usual "ascii text" display, just re-run the BIOGRAPH program and hit the F9 menu choice. Alternatively, use the MS-DOS

"mode" command by typing:

mode bw80 ↩

or

mode mono ↩

F10 = display this menu

This just restores the main menu. As with the F9 pick, the F10 pick can be made from any of the secondary menus, even though it is not displayed as an option. The intent is to provide an easy route back to the main menu.

5.2 Navigating the menu tree

We now describe the actions invoked by choices in each submenu and their submenus. This is for completeness only. It is easier, faster, and more fun to find out what this program does by interactive experimentation than by continuing to read this manual. If you get stuck, need more information, or if a menu choice causes unexpected results, the details below will clarify things. Use the index and table of contents to locate in this manual the menu/submenu options you have trouble understanding.

Each menu, except for the main one, is available in either terse or verbose form. The verbose form slows the response of the program because writing text is slow. However, the verbose form describes each option sufficiently to evoke accurate guesses about what will happen if you pick that option. The terse form gives minimal, thus fast, reminders. You can toggle between terse and verbose forms from any menu[5] at any time using the '!' key.

As you go to deeper menu levels, eventually the menus will be replaced by explicit questions that you must answer by typing numbers or 'y', for "yes", or 'n', for "no"[6], or text names of parameters, etc. These questions are their own descriptions; most are not mentioned in this manual. When you are prompted to enter a number, you *must* enter

[5]except the main menu

[6]Actually any character besides a 'y' or 'Y' is interpreted as a "no" response.

a number; if you type any character other than $\{0123456789\ + -.eE\}$ you will be asked to retype the number.

Alternatively, some menu choices lead to instructions to move a cursor. At this level, you must move the cursor or accept the present cursor position by hitting the space bar. The function keys and all keys not mentioned in the cursor moving instructions will be ignored until you hit the space bar.

For all the secondary menus, the character within square brackets '[ ]' in the menu's right margin shows the key you hit to pick the option. We use the notation "F7≻[g]" to indicate the choice made by striking the 'g' key when the F7 submenu is displayed on the screen.

As discussed on page 21, you can tell the BIOGRAPH program to read files from, and write files to, any subdirectory of any disk drive installed in your computer. The first line of every menu tells you the current path to files. Every menu, except for the main menu, includes the [\] option to change this path, and you can change the path at any time.

5.3 The F1 menu

Here is the secondary menu called by the F1 key from the main menu.

– – –{ current path to files = c:\biograph\ }– – –
[m]–>[m]ake graph from most recent integration data
[q]–>[q]uit; clarify screen
[c]–>add [c]aption(s) to picture <<F1 MENU>>
[s]–>[s]ave picture in a named disk file
[R]–>[R]etrieve picture previously saved in a disk file
[F]–>send [F]orm feed signal to printer
[d]–>display [d]isk directories (& erase files)
[!]–>switch to terse menus and prompts
[\]–> change path to files
selection–>

The [s] and [R] options are self explanatory. To save picture files you must have sufficient free disk storage available; each picture file uses 16,384 bytes. All picture files created or re-read have the **.pic** tag automatically appended to their names. Several such *.*pic* files have been supplied on your diskette. Use the [d]≻[p] option to see their names. Use the [R] option to see samples of the plots this program makes. Before the program replaces an existing picture file with a new version, you will be asked to confirm your intention to do that.

The [c] option initiates a self explanatory dialog in which you move a cursor to any desired position on the screen, then type a caption. You are then asked to confirm that you wish to keep that caption as a permanent part of the graph. Note that captions consisting of blanks can be used to erase undesirable parts of a picture.

F1≻[m]–> [m]ake graph from most recent integration data

Each time a solution is computed, sampled values of each dependent variable component are stored in a large (65 K byte) memory array (even for components not presently scheduled to be graphed). Most of the F5 menu options that trigger integration draw the graph immediately after integrating. If you decide that another graphing style would be preferable, or if you wish to see other components graphed, you can use the F3 menu to specify a new plotting style. Then use this [m] choice to replot the most recently computed solution data, possibly after using the F2 main menu choice to erase the screen. Only the data from the most recent integration is available.

F1≻[q]–> [q]uit; clarify screen

This just purges all menu text from the screen so you can see a "pure" picture and possibly make a hard copy on the printer.

F1≻[q]–> send [F]orm feed signal to printer

This sends a form feed character to the line printer[7] to make the printer move to the start of a new page. This is useful prior to copying the screen to the printer to ensure that each plot starts at the beginning of

[7] an ascii code 12 to the MS-DOS logical device PRN:

a blank page. Use this option only if you have a printer connected and powered on.

F1≻[d]–> display [d]isk directories (& erase files)

Use this to scan disk drives for *∗.pic* files for use by the [R] option. The next section describes the tertiary menu called by this choice.

5.4 Disk file directory menu

The following menu is accessible from the F1, F7, and F8 menus. Using it, you can print directories of file names in the subdirectory of the disk drive specified by the current path to files[8], and you can erase unwanted files. When you make the F1≻[d] pick, you will see this menu:

– – –{ current path to files = c:\biograph\ }– – –
[u]–>[u]ser parameter (∗.par) file name list
[p]–>[p]icture (∗.pic) file name list
[k]–>[k]ey (∗.key) playback file name list
[a]–>[a]ll file names listed
[E]–>[E]rase a file
[q]–>[q]uit task
[!]–>switch to terse menus and prompts
[\]–> change path to files
selection–>

The [u] option displays names of *∗.par* files containing user parameter values, plot style, integrator tolerance, data defining problems and their cases, etc. These files are written and read from the F7 menu.

The [p] option displays all *∗.pic* files that are accessible from the current path to files. Recovery of such pictures, from the F1 menu, is instantaneous.

The [k] option shows names of *∗.key* player piano files containing exact records of the keystroke input stream during a session. Replayed,

[8]The current path to files is always displayed on the screen as the first line of every menu except for the main menu. Instructions for changing this path to files appear on page 21.

these reconstruct a picture, allowing you to see which commands cause which actions. Use the F8 menu to create or replay such *.*key* files.

The [a] option shows you the name of all files on the active drive. If the listing fills more than one screen, use the 'Ctrl' 's' key to halt the listing temporarily, and the 'Ctrl' 'q' key to restart it.

The [E] option allows you to erase files from within the BIOGRAPH program. When asked to type the name of the file, you must type the entire name including the tag (if any). Before the file is erased you will be asked to confirm that irrevocable action. Alternatively, you can erase unwanted files after terminating the program by using this MS-DOS command

erase filename.tag ↩

where "filename.tag' stands for the name of the file to be erased.

No option in this submenu will function unless the MS–DOS file called *command.com* was on the disk drive/subdirectory from which you initiated the BIOGRAPH program.

5.5 The F3 menu

Use this menu to define the style in which graphs are made. After you have computed a solution, you can experiment to find the best style for displaying it. Use this menu to try a new style. Use the F2 option to erase the old plot, then use the F1≻[m] option to draw the old solution in the new style, and continue this procedure until you discover the style you prefer. The F6≻[e] option will tell you the maximum and minimum values attained by the dependent variables in the most recently computed solution, so you can specify appropriate scaling for the plots. Here is a facsimile of the F3 menu.

- - -{ current path to files = c:\biograph\ }- - -
[w]->change the number of [w]indows in the plot <<F3 MENU>>
[h]->specify [h]orizontal scale (same for all windows)
[H]->define [H]orizontal variable (same for all windows)
[#]->specify frequency of sampled values stored during integration.
[p]->specify variables to [p]lot, & their vertical scales and windows
[b]->specify a window's [b]orders
[m]->re-define specify window [m]ap
[W]->specify [W]indow to which to move a variable to be plotted
[q]->[q]uit task
[!]->switch to terse menus and prompts
[\]-> change path to files
selection->

Details of the F3 menu

F3≻[w]-> change the number of [w]indows in the plot

The graphics screen can be partitioned into 1, 2, 3, or 4 horizontal plot strips or *"windows"*. The horizontal scale is common to all windows. Each window can have multiple independent vertical scaling of curves plotted in it. Each of the windows on the screen has the same physical height regardless of scaling. The [w] option asks you how many windows you want.

If, using this option, you prescribe fewer windows on the screen than are scheduled to be filled by the [p] option, you do not annihilate any windows, you just make them invisible. You can see what's in them later by increasing the number of windows.

F3≻[h]-> specify [h]orizontal scale (same for all windows)

This prompts you to enter the minimum and maximum values on the horizontal scale. These values correspond to the left and right edges of each plot window. You can use this option to examine any part of a solution; parts of a solution lying outside the specified min and max values are not plotted, but do no harm.

F3≻[H]–> define [H]orizontal variable (same for all windows)

You can plot the independent variable (t) on the horizontal axis of all plot windows if you answer with 0 (zero) the query triggered by this option. Alternatively, you can plot any dependent variable component along the horizontal axis by typing the component's number in response to the query. Use this latter scheme to make phase portraits.

F3≻[#]–> specify frequency of sampled values stored during integration.

This option is not available for difference equations. As the solution of a differential equation system is computed, sampled values of each dependent variable component are stored in a temporary data disk file. Use this option to specify how many samples are stored. The number of samples cannot exceed 2000.

There are two qualitatively different ways to specify the sampling frequency. One is to take samples at equal increments in the independent variable. Another is to employ "equi–displacement sampling".

> ↝ Though simple to set up, sampling at equal increments of the independent variable has a serious disadvantage for some differential equation systems. Often most of the variation of the dependent variables occurs during short "bursts" of activity between which no variable changes much. In order to "catch" the interesting behavior of such a solution, the sampling frequency would have to be set so that four or five samples were recorded during a "burst". When equal time samples are taken, that may require tens of thousands of samples to be recorded during the long times between bursts, while nothing interesting is happening. Taking tens of thousands of samples is impractical because it is slow and would use too much memory. For this reason, the "equi–displacement sampling" option described below is provided.

If you answer the program's query about how many samples to record by entering a positive integer, you specify that that many samples, at equal increments of the independent variable, are to be saved.

You initiate "equi-displacement sampling" if you answer the query about how many samples to record typing
0 ↩

> ⇝ You will be asked to enter a separate increment for each dependent variable component. During integration, a sample will be recorded whenever any one (or more) of the dependent variable components changes by more then the increment you specified for it. Each variable must have its own increment because, for some ODE systems, the first component may vary between 0.0 and 1.0, say, while another component varies between 0.0 and 1000. This method of scheduling sampling is superior to equal time sampling, but it involves more work to set up; you have to specify several increments intelligently. Also, since no more than 2000 samples are allowed during a single integration process, the integration may be terminated before the final time you specified is attained because all available memory for sample points was used up.

Saving and retrieving samples takes some time so the more values stored, the slower the program's response will be, but the smoother the plots will be.

We store sampled values of a solution, rather than just plotting the curves as the solution is computed, so that you can experiment with plotting style without the (sometimes slow) recomputation of a system's solution.

F3≻[p]–> specify variables to [p]lot & their vertical scales and windows

You can plot any dependent variable component in any active plot window, with arbitrary vertical scaling. You *must* use the [p] option or nothing will be drawn. The [p] option initiates a dialog in which you declare which component(s) to plot, what line style to use, which window holds the plot, and how the vertical scaling is defined. The queries are self explanatory. The lower numbered window panels appear lower on the screen. Window number 1 is at the bottom and window 2 is on top of window 1, etc.

When asked what line style to use to draw a plot, if you do not wish to use the default solid line style, then you can specify arbitrary gap–dash line styles.

↝ One special line style is no lines at all. When asked to specify the line style code, answer 0 (zero) to specify that no line segments are to be used to connect points on the plot. Just single pixel dots will then be used. This feature is useful when displaying the analog of phase portraits for systems of finite difference equations.

F3≻[b]–> specify a window's [b]orders

Each plot window can be framed by a rectangular border drawn in an arbitrary line style, and having any number of equally spaced tic marks on its vertical and horizontal axes. The [b] option initiates a self-explanatory dialog to make the appropriate choices.

F3≻[m]–> specify [w]indow map

This option lets you re–direct plot window i to actually occupy window number j, for $i = 1, 2, 3, 4$ and for $j = 1, 2, 3, 4$. This is useful, for example, for overlaying four windows in the same physical space. You do that by using the [w] option to specify just one window, and this option to map windows $1, 2, 3$ and 4 all into window 1. This gives each curve the full vertical height available on the screen. If the plots are so intricately interwoven that they cannot be distinguished, you can separate them by using this option to map window 1 to 1, 2 to 2, 3 to 3, and 4 to 4. Then use the [w] option to specify four (now separate) windows. Experiment to see how this feature works.

F3≻[W]–> specify [W]indow to which to move a variable to be plotted

This option initiates a sequence of self-explanatory queries to facilitate moving variables scheduled for plotting in one window to another window. This is faster than using the [p] option to accomplish the same result.

5.6 The F4 menu

Here is the secondary menu called up by the F4 key from the main menu. You use it to modify or inspect the initial conditions (abbreviated *"IC's"*). Most of its options are self-explanatory.

– – –{ current path to files = c:\biograph\ }– – –

[d]->[d]isplay current initial conditions <<F4 MENU>>

[c]->change one [c]omponent's initial values

[f]->set IC's = [f]inal dependent var. values from last integration

[p]->PHASE PORTRAIT ONLY: get IC [p]oint from cursor

[k]->[k]ey in new initial conditions for all components

[o]->restore [o]ld initial conditions from previous integration

[q]->IC's ok; [q]uit task

[!]->switch to terse menus and prompts

[\]-> change path to files

selection->

The [k] option displays the current initial condition for each dependent variable component, then asks you to type in a new value. Use [c] if you want to change only one component's initial condition. Use [k] to change them all. The F4≻[f] option is useful if, after integrating one segment of a trajectory, you wish to continue the integration further, starting from the point where the last segment ended. Various actions described in this section and the next change the current initial conditions. In case you make an error, the F4≻[o] option is useful to restore the initial conditions from which the most recently computed trajectory started.

5.7 The F5 menu

Here is the secondary menu called up by the F5 key from the main menu. This is the menu you use to compute solutions of differential or difference equations, to construct phase portraits, or to hunt for steady state points (critical points).

- - -{ current path to files = c:\biograph\ }- - -
[d]->[d]isplay initial and final times
[I]->new [I]nitial time <<F5 MENU>>
[f]->new [f]inal time
[i]->[i]ntegrate, save data, but don't plot yet
[m]->integrate, save data, and [m]ake the plot
[t]->integrate for specified [t]ime segment, then update IC's
[a]->PHASE PORTRAIT ONLY : draw one [a]rrowhead at cursor
[A]->PHASE PORTRAIT ONLY : draw a grid of flow [A]RROWS
[p]->PHASE PORTRAIT ONLY : get new IC [p]oint using cursor
[n]->PHASE PORTRAIT ONLY : find/draw [n]ullclines
[c]->seek [c]ritical or steady-state point.
[o]->recover [o]ld initial conditions
[k]->[k]ey in new initial conditions
[q]->[q]uit task
[!]->switch to terse menus and prompts
[\]-> change path to files
selection->

Details of the F5 menu

The [d] option shows the independent variable interval for which solutions will be computed by the next pick of the [i] or [m] options. The independent variable is called "time". Use the [*I*] and [f] options to change the endpoints of this interval. Make the final time less than the initial time to integrate backwards. Do not try this for finite difference equation systems because backward calculation generally makes no sense in that context.

The initial conditions, changed using the [k] or [p] options, or from the F4 menu, are imposed at the "initial time" shown by the [d] pick.

Both the [i] and [m] options trigger numerical integration of the differential (or difference) equation system. In either case, the initial conditions remain unchanged. The [m] and [i] options differ only in that [i] supresses plotting. That is useful if you do not want to contaminate a plot in progress before you see (using the F6 menu's choices) "where the new solution went".

The [t] option asks you to specify a final time to which to integrate.

Type in a negative value to integrate backwards. The [t] option then triggers integration and drawing, and finally sets the new initial conditions equal to the old final values attained by the dependent variables. The total successful integration time is added to both the initial and final time limits to update those values. The F5≻[t] option is equivalent to using the F5≻[m] option followed by the F4≻[f] option.

> ⤳ This is useful when constructing phase portraits, when one does not know how long to integrate to obtain a desired portion of a trajectory. (Try relatively short times, building the whole trajectory from short segments.) It is also useful when one wants to deposit arrowheads along trajectories. Do that by alternating between the [t] and [a] options. It is efficient to use the F8≻[k] option to create a **.key* playback file that gives a sequence of F5≻[t] options (with, say $t = 0.5$ for each) alternating with F5≻[a] options. Each time the cursor is put at a start position, playback of that file will compute and draw a trajectory with arrowheads spaced along it at equal time units. The file called *arrows.key* does this, with ten arrowhead segments.

When you make the [i], [m], or [t] picks, a progress report will appear on the screen that tells you how the integration algorithm is performing. Writing out all the numbers in this report takes time and thus slows the integration process significantly. You can toggle the reports off and on with the 'r' key. When you suppress the reports with the 'r' key to speed up the solution process, the letter "**W**" in the word "**Working**" blinks on the screen at a frequency that indicates the success rate of the integrator routine. Rapid blinking means rapid progress; slow blinking means the integrator is working hard to re-adjust the integration time step.

While integration is in progress, a screen message will tell you that you can abort the calculation by hitting the 'q' key. Whatever portion of the solution that has been computed will then be drawn.

> ⤳ You can recover the most recent "old" initial conditions using the F5≻[o] or the F4≻[o] options. This is generally useful, but especially so in phase portrait construction, to draw both "halves" of a trajectory emanating from a given point – as t runs forward, then backward:
>
> - put the graphics cursor at a desired point;
> - make the [a] pick, to draw one arrowhead;
> - make the [t] pick and specify a positive time, to create the "forward half" of the trajectory;

- make the [o] pick to get back to the original point;
- make the [t] pick and specify a negative time, to create the "backward half" of the same trajectory.

The [A] option fills a phase plane with a grid of flow arrows, after asking you how closely to space them. This is useful to get an initial impression of the flow and to visualize where steady state points might lie, etc. You will probably have to erase the screen and start over, however, to get a "clean" picture.

If you are making a phase portrait and want to specify new initial conditions using the graphic cursor, make the [p] pick. You will then see instructions for moving the cursor. Refer to the last subsection of this chapter on page 118. If you are in the terse menu mode, these instructions will not be displayed, but you can call them up by hitting the 'h' (for help) key.

F5≻[n]–> PHASE PORTRAIT ONLY : find/draw [n]ullclines

If you are making a two dimensional phase portrait, or its analog for a system of two finite difference equations, this feature lets you find and display the nullclines. See MDP for the definition and use of nullclines.

↝ The algorithm for finding nullclines involves the automatic construction of a new system of differential equations, built from the original ones, having the nullclines of the original system as stable invariant manifolds. Solutions to this new system "seek" those nullclines. Once one is found, it is drawn as the solution runs along it. This algorithm is presented in Phillips and Odell, (1984)*.

Hunting for nullclines is an interactive process. Nullclines can have many disjoint branches, and the algorithm for finding them seeks the one nearest the initial conditions given by the cursor. Thus, for complicated phase portraits, you may have to begin from several different initial conditions. To see a demonstration of this procedure, play back the file called *acubic.key*. Do this by picking the [P] option from the F8 menu, then typing the file name: **acubic**.

Here is the message and tertiary menu displayed when you make the [n] choice from the F5 menu:

The part(s) of the nullclines found depends on where
the cursor is when this process begins...

You can abort the nullcline hunt at any
time by pressing the 'q' key (possibly several times).

Hit the space bar to continue.

space bar ↩

- - -{ current path to files = c:\biograph\ }- - -
[x]->seek/draw (vertical) [x]-nullcline
[y]->seek/draw (horizontal) [y]-nullcline <<F5 sub-menu>>
[b]->seek/draw [b]oth nullclines
[a]->change [a]ccuracy with which nullclines are located
[#]->change [#]of nullcline markers
[p]->get new search start [p]oint using cursor
[q]->[q]uit task
[!]->switch to terse menus and prompts
[\]-> change path to files
selection->

As the (sometimes slow) calculation proceeds, the nullclines will be marked by short vertical and horizontal bars. The [#] option determines their spacing. As discussed in the text, critical points or steady state points occur precisely where the two nullclines intersect. The F5≻[c] option provides an independent procedure for finding steady state points (critical points). Indeed, it may well be advisable first to find a steady–state point and then to use this point as the initial condition for the nullcline search.

↝ You control the accuracy to which nullclines are found using the F5≻[n]≻[a] option. Smaller error tolerance specifications made in response to the query triggered by the [a] pick mean more accurate (and slower) nullcline location. Note that there is not much point in locating nullclines more accurately than can be displayed on your graphics screen.

F5≻[c]–> seek [c]ritical or steady–state point.

This option triggers a search for steady state points (or critical points). The search begins at the present initial conditions. There is no limit to the number of differential equations in the system for this option.

> ↝ A variant of Newton's method is employed. The Jacobian matrix [9] of the right hand side of the differential or difference equation system is approximated numerically by finite difference quotients. If the order of the system is higher than three, the program merely displays the numerical coordinates of a steady state point if it finds one. For second and third order systems, the jacobian matrix evaluated at the steady state point and the eigenvalues are computed and typed on the screen. The "characteristic equation" typed on the screen is equivalent to *(MDP:-A4.11, p246)*; see also *(MDP:A4.21, p248)*. For second order systems, any steady state point found is classified (saddle point, node, spiral etc.). When in the phase portrait mode, the graphics cursor jumps around, showing you the points visited by Newton's method during the search. A heuristic modification to Newton's method stabilizes (and retards) the first eight iterations, so these may make less progress than pure Isaac Newton would have, but by being less reckless, they are more dependable than pure Isaac. The iterative search for a steady–state point terminates successfully when the "function norm" becomes sufficiently small. The function norm is the maximum absolute value of any component of the right hand side, $\vec{f}(\vec{x})$.

5.8 The F6 menu

Using this menu's options, you can display all the current information stored by the program concerning a particular problem/case/solution. The information is listed out on the screen. In addition, if you enable "hardcopy printing" and you have a printer hooked up, then the same information will be printed. One purpose of this feature is to provide you with paper copies of all important parameter values, plot style, etc. associated with a problem you have solved and plotted on the printer. Here is the secondary menu called up by the F6 key from the main menu.

---

[9] The Jacobian matrix is essentially the coefficients of *(MDP:A5.14, p254.)*

```
---{ current path to files = c:\biograph\ }---
[h]->toggle [h]ardcopy printing on/off
[e]->show [e]xtremal values for dep. var. in last integration
[n]->show [n]umber of equations in system
[u]->show values of [u]ser-defined parameters
[c]->show values of integrator [c]ontrol parameters     <<F6 MENU>>
[p]->show [p]erformance during last integration
[i]->show [i]nitial conditions
[t]->show initial and final values of [t]=independent variable
[f]->list [f]inal values of dependent variables
[v]->display [v]ariables plotted, and their windows
[A]->report [A]ll (useful for hardcopy listing)
[C]->list file defining this [C]ase
[q]->[q]uit task
[!]->switch to terse menus and prompts
[\]-> change path to files
selection->
```

Details of the F6 menu

Most options are self-explanatory. We provide a few notes, however.

F6≻[h]-> toggle [h]ardcopy printing on/off

Do not use this option if you have no printer. When printing is disabled, the [h] option turns it on; when printing is on, [h] turns it off. In any case, printing is always turned off when you quit [q] the F6 menu. If the power switch of your printer is turned off, or if the printer is not connected, you will get a message from the MS-DOS operating system asking whether to retry, abort, or ignore the problem. Make sure your printer is hooked up, plug it in, turn it on, and then pick the "retry" option.

While hardcopy printing is enabled, all alphanumeric information displayed on the screen is typed verbatim on your printer. This method of printing is *much* faster than using the 'shift' 'PrtSc' key to obtain hard copy of graphs over which the program has displayed alphanumeric information.

F6≻[p]–> show [p]erformance during last integration

This tells you whether the Adams-Moulton variable order predictor–corrector algorithm performed efficiently during the most recent integration process. Ideally the average number of function evaluations per step should be small – say 2 or 3. If, according to the report, the number of step size cuts was more than ten percent of the number of function evaluations, you should specify a smaller maximum allowable integration step size using the F7≻[c] menu choice.

F6≻[A]–> report [A]ll (useful for hardcopy listing)

This triggers a full dump of everything the program knows about the current problem/case/solution. It is the same as making all of the other picks in sequence. The information will scroll past on the screen faster than you can read it. Use the 'Ctrl' 's' key to halt the listing temporarily, and the 'Ctrl' 'q' key to restart it.

F6≻[C]–> list file defining this [C]ase

This option will display an exact copy of the *.*par* file for the current problem/case as it would be written to disk by the F7≻[S] option discussed in the next subsection.

> ↝ This file is standard ascii text and can be edited. The position of data within each line in the *.*par* files is important. The (vertical) order of lines in the file is not important.

5.9 The F7 menu

The secondary menu called up by the F7 key from the main menu allows you to read and write the *.*par* disk files that store information on parameter values, plotting style, etc. Also, you can change the values of parameters that determine integrator performance and that determine the right hand sides of differential and difference equations. The F7 menu looks like this:

- - -{ current path to files = c:\biograph\ }- - -
[u]->display/change [u]ser-defined parameters
[c]->display/change integrator [c]ontrol parameters
[i]->[i]dentify present problem-case; initialize
[S]->[S]ave present problem-case on disk
[g]->[g]et previously stored problem-case from disk *<<F7 MENU>>*
[m]->[m]ake/compute/draw solutions - - - >F5 menu
[d]->display [d]isk directories (& erase files)
[q]->[q]uit task
[!]->switch to terse menus and prompts
[\]-> change path to files
selection->

Details of the F7 menu

F7≻[u]-> display/change [u]ser-defined parameters

Each **.par* file defines the text names of parameters occurring in the right hand side of the difference or differential equation. The [u] option allows you to change the *numerical values* of these parameters, *not* the names[10].

In every case, a parameter called *select* appears. Its value determines which problem's equations are evaluated. The most recently read **.par* file determines which named parameters are loaded. The parameters must match the equations for proper operation of the program.

> ↝ The values of the *select* parameter corresponding to each problem are given in chapters 3 and 4. In most cases you can change to a new problem only by reading in a new **.par* file, using the [g] option of the F7 menu. In two cases, however, you can change problems just by changing the value of the *select* parameter. You can exchange *select* = 2 with *select* = 3 to alternate between the second and third order versions of the cAMP relay–oscillator model. You can exchange *select* = 7 with *select* = 8 to alternate between third and second order versions of the enzyme–substrate problem. In each of these cases, the two– and three–dimensional versions of the problem have exactly the same parameters.

[10] Use a text editor to change the parameter names by modifying the **.par* files.

The [u] pick initiates a dialog, of which the following is an example, showing how the value of a parameter is changed in the differential equation for the third order cAMP relay–oscillator problem (*select* = 3).

<< User-defined parameter values: >>

sigma=12.00000 q=100.00000 kt= 4

k=4 h=10.00000 L=1000000.00000

d=0 s=0 select=3

Name of parameter to change (just return for none)

sigma ↩

Currently, sigma= 12.000000

Enter new value (floating point)

11.345000 ↩

<< User-defined parameter values: >>

sigma= 11.34500 q= 100.00000 kt= 4

k= = 4 h= 10.00000 L=1000000.00000

d=0 s=0 select=3

Name of parameter to change (just return for none)

↩

F7≻[c]–> display/change integrator [c]ontrol parameters

You can and should control the performance of the ODE integration algorithm by tuning the five parameters that the [c] option makes available. Here is an example of the dialog triggered by this option:

Current...

<< Integrator control parameter values: >>

kmax=6, maxitt=3, etol= 0.00010000

hmin= 0.00000100, and hmax = 2.000000

Type name of parameter to change (return if no further changes)

hmax ↩

New hmax (floating point, > hmin)?

5.000000E-1 ↩

↝ For help in understanding the technical terms below, you should refer to any modern text in numerical analysis that discusses ODE initial value problem integration. See, for example, Burden and Faires (1985)*.

The integration algorithm uses paired explicit predictor and implicit corrector multistep formulae to compute the value of the solution at the present time $+h$, where h is the variable integration time step. There are six pairs of such predictor–corrector formulae built into the program, of orders $1, 2, ..., 6$. An order k pair of formulae approximates the derivative of the solution using a degree k polynomial, so the higher k is the less local truncation error is made for a given step size, h. The algorithm automatically tries to use the highest order method you permit. *kmax* is the maximum allowable order, and must be set between 1 and 6. For some very stiff differential equations[11] lower maximum order means more stable behavior of the integration algorithm; even restricting to $kmax = 1$ may be appropriate for very stiff systems.

At the beginning of each integration (initiated by the [m], [i], or [t] options of the F5 menu) the integration scheme starts out using a first order predictor–corrector pair of formulae. The truncation error made is of order $O(h)$, where h is the length of the integration time step. To keep that truncation error below the specified error tolerance, *etol*, it is usually necessary to make very small integration steps. After one or more "tiny" steps, which create the "history" that higher order multistep methods require, the integration algorithm switches to a second order scheme and hence makes truncation errors of order $O(h^2)$, so that larger steps are possible while meeting the same *etol* truncation error criterion. After some more integration steps, the integrator algorithm decides whether to increase the step size, h, or switch to a third order predictor–corrector pair. This process of increasing the length of the time step, h, and/or the order of the predictor–corrector pair continues until the highest order method (up to sixth) permitted by the user–specified value of *kmax* is reached, and until

[11] The equation system is "stiff" where one or more of the right hand sides vary greatly as the dependent variables change. Formally this means that the $d\vec{f}/d\vec{y}$ jacobian matrix is 'large' in norm.

the time step reaches the minimum of:

- *hmax* specified by the user (via the F7≻[c] menu option)
- the largest step size for which the fixed point iteration solution of the corrector formula converges within *maxitt* iterations
- the largest step size for which the truncation error is less than *etol*.
- $$t_{SAMPLE} = \frac{|\mathbf{tmax} - \mathbf{tmin}|}{\textbf{number of points to be stored for plotting}}$$

The parameters in the first three items above are set using the F7≻[c] menu choices, and the number of plotting samples stored is determined using the F3≻[#] option.

Heuristic aspects of our algorithm's strategy for manipulating the order of the predictor-corrector scheme, the integration time step size, etc. are not discussed in this manual; their only justification is that they succeed.

The start up procedure can take some time. If you watch the integration progress reports, you will notice that it may take quite a few seconds before the first sample point is saved for plotting.

The number of sample points you elect to save for plotting influences the maximum time step allowed only when you select equal time increment sampling in the F3≻[#] option. Note also that if you save infrequent samples of a solution, then you will get jagged plots — no matter how accurately the solution was computed. More often than not for the problems treated by the BIOGRAPH program, the time interval between plotted points in a "nice" plot is substantially smaller than the time step the integration algorithm *could make* and still keep the truncation error below *etol*. The best way to avoid jagged plots (and to remove the influence of the sampling interval from the maximum time step used) is to select "equi-displacement sampling" from the F3≻[#] option.

The implicit corrector formulae, which are systems of nonlinear algebraic equations, are solved by fixed point iteration[12]. If more than *maxitt* iterations are needed by the fixed point iteration scheme to attain convergence of the solution to the corrector formulae, then the time step, h, is cut. As $|h|$ gets smaller,

[12] not by Newton's method as in Gear's algorithm

fewer and fewer fixed point iterations are needed to get convergence. $maxitt = 3$ is usually best, but you should experiment with maxitt to speed up the integration process if it becomes suspiciously slow.

The integration algorithm automatically decreases the step size, h, so that the local truncation error made on each step never exceeds *etol*. The step size is increased whenever the local truncation error falls below $etol/10$. In this way the integration process can use large time steps through regions where the solution is changing slowly, but can sense regions of abrupt variation of the solution and use appropriately smaller time steps. You get more accurate solutions (at the cost of longer computation time) by making *etol* smaller. You may want to use a larger *etol* value to compute rough approximations quickly while you are exploring new parameter values, then cut *etol* for final "production" runs. $etol = 0.001$ is a good base value from which to explore changes in algorithm behavior as *etol* varies.

The user constrains the local truncation error by picking a value for *etol*. The next several paragraphs explain the relationship between the local truncation error *made at each step* during an integration and the total error made during an integration of length T, which requires many steps.

In rough terms, the following error tolerance is controlled by *etol* at each integration step. Let $u(t)$ be any one of several variables in the system of differential equations that is being solved. The BIOGRAPH program adjusts the time increment h for each integration step so that

$$\frac{\left|u^{exact} - u^{approximate}\right|}{|h| \; (1 + |u^{approximate}|)} < etol$$

It follows from this above formula that if u is large compared to unity then the relative error is controlled:

$$\left|\frac{u^{exact} - u^{approximate}}{u^{approximate}}\right| < etol \cdot h \quad .$$

If u is small compared to unity then control is enforced on the absolute error:

$$\left|u^{exact} - u^{approximate}\right| < etol \cdot h \quad .$$

When the above constraint is maintained at each time step, we can be sure that, after an integration of total length T, we will have the total error bounded as follows.

$$\left|u(t_0+T)^{exact} - u(t_0+T)^{approximate}\right| < etol \cdot F(T) \quad .$$

where $F(T)$ is a function of T independent of h or the number of integration steps taken. Thus the error can be reduced by diminishing $etol$, no matter how many time steps are made.

We now supply more complete technical detail concerning the error analysis. When the integration scheme is using a kth order predictor-corrector pair, the history of the solution

$$\vec{y}_{j-k+1}, \ldots, \vec{y}_{j-1}, \vec{y}_j$$

is saved. Here $\vec{y}_j$ is the value of the solution at time t_j, with

$$t_{j+1} - t_j \equiv h(j) \equiv \text{ integration step size at step j.}$$

Suppose for the moment that $\vec{y}_{j-k+1}, \ldots, \vec{y}_{j-1}, \vec{y}_j$ were *exact* values of the solution. From this information, an approximation to the solution at time t_{j+1} is computed, as stated above, by an implicit corrector formula solved by fixed point iteration. Call this new solution $\vec{y}^{\,approx}_{j+1}$.

Let $\vec{y}^{\,exact}_{j+1}$ denote the exact value of the solution at time $t_{j+1} = t_j + h(j)$. Even if the historical values from which $\vec{y}^{\,approx}$ is computed are exact, $\vec{y}^{\,approx}_{j+1}$ will differ from $\vec{y}^{\,exact}_{j+1}$. The conventional *local truncation error of the ODE integration scheme* measures the difference as follows.

$$\tau_{h(j)} \equiv \frac{\left\|\vec{y}^{\,exact}_{j+1} - \vec{y}^{\,approx}_{j+1}\right\|}{|h(j)|} .$$

Here the generalized magnitude or *norm* $\|\ \|$ is defined by

$$\left\|\vec{y}\right\| \equiv \max_{i=1}^{n} |i^{th}\, component\ of\ \vec{y}|$$

Note that the local truncation error $\tau_{h(j)}$ has an $h(j)$ in the denominator. Keeping such an expression less than some specified tolerance, $etol$, is mandatory for the following reason. Let us restrict ourselves for simplicity to the case $h(j) \equiv h =$ constant. Then when integrating for a total time of T, we make T/h integration steps. We thus naively expect that

$$\sum_{j=1}^{T/h} \frac{\left\|\vec{y}^{\,exact}_j - \vec{y}^{\,approx}_j\right\|}{h} < \frac{T}{h} \cdot etol \quad .$$

This would imply that

$$\text{total error } \equiv \sum_{j=1}^{T/h} \left\| y_j^{exact} - y_j^{approx} \right\| < etol \cdot T \quad .$$

The previous arguments were heuristic. They do not take account of the fact that because each $\vec{y}_{j+1}^{approx}$ is computed from erroneous historical values (not exact ones as assumed above), every error made can be compounded. The rigorous theorem assumes the existence of a Lipschitz constant K, such that:

$$\left\| \vec{f}(t, \vec{y}) - \vec{f}(t, \vec{z}) \right\| \leq K \left\| \vec{y} - \vec{z} \right\|$$

for every t, $\vec{y}$ and $\vec{z}$ in the domain of interest. Then it can be shown that

$$\left\| \vec{y}(t_0 + T)^{exact} - \vec{y}(t_0 + T)^{approx} \right\| \leq \frac{\tau_h^* \cdot e^{K|T - t_0|}}{K}$$

where τ_h^* is the largest of the $\tau_{h(j)}$ and t_0 is the time at which integration commences.

To compensate for the fact that different components of the $\vec{y}(t)$ vector could have very different magnitudes, BIOGRAPH actually controls this normalized version of local truncation error:

$$\hat{\tau}_{h(j)} \equiv \max_{i=1}^{n} \frac{\left| \left(\vec{y}_{j+1}^{exact} - \vec{y}_{j+1}^{approx} \right)_i \right|}{h \left(1 + \left| \left(\vec{y}_{j+1}^{approx} \right)_i \right| \right)} \quad ,$$

where $\left(\vec{z}\right)_i$ denotes the ith component of the n-vector $\vec{z}$. Keeping

$$\hat{\tau}_{h(j)} < etol$$

is a still more stringent suppression of local truncation error than is the conventional measure described above. The error estimates bound the error by an exponentially increasing function of $|T - t_0|$. In favorable cases the error may be much less than this bound[13].

[13] Our discussion of truncation error does not include the effects of round-off, the error introduced by the fact that the computer calculates with double precision numbers rounded off to 16 decimal places. Thus difference equations are *not* solved "exactly". All users of computers should be aware that round-off errors can accumulate, for some sensitive problems very rapidly, to render meaningless the calculated solutions. See N. Liron's Appendix 5 in Segel (1984) for a brief introduction to basic issues in numerical analysis, in a biological context.

hmin and *hmax* are the minimum and maximum allowable time steps the integration algorithm is allowed to make. The time step, h, can be positive or negative; the following condition is enforced:

$$hmin \leq |h| \leq hmax \quad .$$

As a differential equation solution trajectory enters regions of increasing stiffness, the integration algorithm cuts the time step to whatever small value is needed to keep the truncation error smaller than *etol* and to have the fixed point iteration converge. Some differential equations have solutions that go to infinity in finite time. If you try to integrate one of those, the integration scheme will cut the time step arbitrarily close to zero, and thus compute infinitely many steps. You can prevent this futile process by specifying an appropriate minimum time step, *hmin*. If the algorithm needs $|h| < hmin$, the integration calculation is terminated, and a message is displayed on the screen.

It is crucial to adjust *hmax* appropriately. If you make *hmax* too large, the integration algorithm may keep trying to boost the time step toward *hmax*, only to find it must immediately cut the time step to achieve the desired *etol* truncation error. The algorithm *must* have this property in order to adjust the time step within wide limits to values that are appropriate for the region being traversed. Both increases and decreases of the time step slow the calculation. Consequently you should reduce *hmax* if you see that too many time step cuts have been made. Both the on-screen progress report and the F6≻[p] option will tell you how often the time step was cut. Too small a value for *hmax* will lead to accurate but very slow calculations.

In a typical situation where *hmax* should be reduced, you will see many cuts and boosts announced on the screen, accompanied by fluctuations in the step size. If you fix *hmax* a bit smaller than the typical step size just prior to a size cut, then cutting and boosting will cease and the integration will run significantly faster. It could be, however, that the problem later becomes less stiff, in which case the recommended strategy prevents a later (beneficial) increase in the step size.

F7≻[i]–> [i]dentify present problem-case

After loading a new problem/case file using the [g] option described below, it is a good idea, but not essential, to invoke this [i] option. It causes a screen display that tells you what problem/case is currently active. Typically you will see references to particular equation numbers on a certain page of MDP. Other messages report which components of the dependent variable corresponds to which variable in the text book. If you have loaded a finite difference equation system instead of a differential equation, a message will tell you that.

For many problems, in addition to printing the above sort of message, this option will check to see if the present parameter values satisfy certain constraints, detailed in the main text book, that are necessary for the model to make physical sense. Finally, consistency checks are made to verify that the value of the select parameter matches the **.par* file currently loaded, and in some cases routines to calculate certain required dimensionless parameters can be initiated.

F7≻[S]–> [S]ave present problem-case on disk

Note the upper case S. At any time you can invoke this option to save in a disk file, whose name you will be asked to supply, a **.par* file that contains all the information needed to define the present state of the program. This includes the current problem/case, all information about plotting style, integration control parameters, initial conditions, initial and final integration times, etc.[14]. If the file name you specify is the name of an existing file, you will be asked whether you want to obliterate the existing file[15] by writing a new one "on top of it".

Never type the *.par* tag in response to the file name query. That is, to read in the file called *2d_camp.par*, type:

2d_camp ↩

↝ When, for the [S] or [g] options, the program asks you to

[14] Whenever you use the F7≻[g] option, described next, to re-read such a file, you can recover the state of the program at the time you saved the **.par* file.

[15] After terminating the BIOGRAPH program with the F9 choice, you can erase files by using either of the MS-DOS commands

erase filename ↩

or

del filename ↩

where "filename" is the name of any file you no longer want. Alternatively, the [d] option calls the disk file inspection submenu that includes the [E] option to erase unwanted files from within the BIOGRAPH program. See page 90.

specify the name of a disk file to be read or written, the current path to files determines which subdirectory of which disk drive is used. That current path is displayed in the first line of every menu and the [\] option of every menu allows you to declare a new current path.

F7≻[g]-> [g]et previously stored problem-case from disk

This is the inverse operation of the F7≻[S] option. To begin to work on one of the differential or difference equation problems that this program treats, you *must* use this option to read a **.par* file that defines the parameter names and values, order of the system, etc. from the disk. Many such files have been supplied for you to read, then modify. You will be asked to type the file name; The current path to files is used; see the preceding paragraph.

F7≻[m]-> [m]ake/compute/draw solutions —> F5 MENU

This is just a short cut route to the F5 menu. It is equivalent to making the F7≻[q] pick to return to the main menu, followed by the F5 pick from that main menu.

F7≻[d]-> display [d]isk directories (& erase files)

Use this option to scan the disk directories for **.par* file names you use with the [g] option above. It invokes the tertiary menu described on page 89.

5.10 The F8 menu

Here is the secondary menu called up by the F8 key from the main menu.

- - -{ current path to files = c:\biograph\ }- - -
[L]-> Initiate session [L]ogging.
[k]-> initiate disk saving of [k]eystrokes
[P]-> [p]layback a previously saved keystroke file
[d]-> display [d]isk directories (& erase files)
[q]-> [q]uit and continue with program <<F8 menu>>
[!]->switch to terse menus and prompts
[\]-> change path to files

This menu allows you to save two different kinds of disk file records of any session with the program. One kind can be replayed later, in a sort of player piano concert, to recreate every step of the original session.

F8≻[L]–> Initiate session [L]ogging .

↝ Make this pick to start saving in a disk file, whose name you are asked to supply, a complete record of everything typed by the program on the display screen and everything you type on the keyboard. This file can be listed on a printer and can be merged into a document prepared with a word processor program. The facsimile copies of menus and the sample user↔program interactions in this manual were produced that way. When you initiate session logging with the [L] option, you will be asked to enter the ascii codes for characters to flag the start of each line typed on the keyboard, and another set of characters to flag the start of each segment of text produced by the program. You can ignore this option, or you can use it to enter special font flags your word processor understands so that, when the session log is printed, the user's part will be typed in a distinct font.

Session log files have a .log tag appended to their names automatically. No such files have been supplied on your diskette.

Session logging is terminated in two ways. If, when you use the F9 function key to quit the whole program, a session log file is open, then it is automatically closed and saved. Alternatively, if you return to the F8 menu while session logging is active, the [P] option in the F8 menu will be replaced by another that explicitly terminates logging.

F8≻[k]–> initiate disk saving of [k]eystrokes

When you select this option, you will be asked to type the name of a disk file in which every key you hit will be recorded. Only the keys you type (not the program's messages) will be stored. The cursor movement arrows and the function key codes will also be stored.

↝ These are not standard ascii codes, so the resulting file may be hard to edit directly unless you have an editor that can manipulate all 256 byte values, but see the next section.

The keystroke saving files generated by this option have a "*.key*" tag automatically appended to their names. If you specify the name of a ∗*.key* file that exists already, you will be asked to confirm that you intend to replace the existing file by the new one.

You should create, then replay ∗*.key* files for sequences of frequently used commands. That is, the [k] and [P] options of the F8 menu provide a command macro facility.

You terminate keystroke saving by returning to the F8 menu. Then, in place of the F8≻[k] option, you will find an F8≻[S] option that stops the keystroke saving function and closes the file. Alternatively, when you use the F9 function key to quit the program, any active keystroke file will be closed and saved.

Editing ∗.key files

↝ Since the ∗*.key* files will contain non–ascii characters for each of the F1—F10 keys and for the arrow keys that move the cursor, ∗*.key* files are not easy to edit using the typical text editor. It is important, however, to be able to edit these files because, otherwise, a single small error made in the middle of recording a long ∗*.key* file would ruin it irrevocably. It would also be impossible to "splice together" several ∗*.key* files into a larger one. If you have an editor program that can operate on files containing non-ascii character codes with values larger than 128, and lines of characters of great length, then you can try to edit "raw" ∗*.key* files.

Alternatively, a program called *keytrans.exe* is included on the distribution diskette. This program will translate any **.key* file into another file that can be edited using any text editor that produces a straight ascii output file. For example, to edit the file called *insect.key*, issue this command:

keytrans insect.key ↩

The *keytrans.exe* program will then create a file called *insect.ked* for you to edit. The '*.ked*' tag identifies editable versions of **.key* files. Every such file begins with several lines that list the mnemonic "tokens" (two– or three–character strings starting with '\') in the *.ked* file that stand for the non-ascii characters in the original **.key* file. These tokens are defined in Table 5.1.

The \ > token signifying a pseudo line break is needed for the following reason. Lines in *.key* files can be arbitrarily long (e.g. 300 cursor move keys in an unbroken sequence). Most text editors forbid very long lines. Thus the *keytrans.exe* program puts line breaks (carriage returns) into the translated *.key* file, if necessary, so that no line of the resulting *.ked* file is longer than 72 characters. Any such line break, *not in the original* .key *file* is flagged by a \ > token just preceding it so that, on reconversion to a *.key* file, that spurious line break can be recognized and deleted. When editing a *.ked* file you can remove or add such extra line breaks, but make sure that the \ > tokens occur *only as the last two characters on a line*. Sometimes text caption strings in *.ked* files are broken in half by \ > line breaks. But they are rejoined when reconverted by *keytrans.exe* to *.key* files.

You can edit the file called *insect.ked* as you wish, but be careful. Every character in it means something, and careless removal of blanks, etc. will result in a useless **.key* file.

After editing the file called *insect.ked*, you recover a useable **.key* file that the BIOGRAPH program can "execute" by using the *keytrans.exe* program again, this time by issuing this command:

keytrans insect.ked ↩

The *keytrans.exe* program will then translate the contents of the file called *insect.ked* back into the proper format for a **.key* file, called *insect.key* (and replace any existing file with the same name with the new version).

The *.ked* and *.key* tags in the file names of the *keytrans* com-

token	key
\N	‘↑’ cursor key; North or up
\S	’↓’ cursor key; South or down
\E	‘→’ cursor key; East or right
\W	‘←’ cursor key; West or left
\ne	cursor key; Northeast, up and right
\nw	cursor key; Northwest, up and left
\se	cursor key; Southeast, down and right
\sw	cursor key; Southwest, down and left
\F1	‘F1’ function key.
\F2	‘F2’ function key.
\F3	‘F3’ function key.
\F4	‘F4’ function key.
\F5	‘F5’ function key.
\F6	‘F6’ function key.
\F7	‘F7’ function key.
\F8	‘F8’ function key.
\F9	‘F9’ function key.
\F0	‘F10’ function key.
\ >	pseudo line break

Table 5.1: Definitions of tokens in .ked files generated by the *keytrans.exe* program.

mand line tell the program which type of file to translate, and thus to create a file of the other type.

Every editable token in *∗.ked* files for non–ascii characters in *∗.key* files begins with the '\' character. The *keytrans.exe* program tries to translate every 'word' following a '\' symbol in a *∗.ked* file into one of the special characters in the *∗.key* files. Therefore, never use '\' symbols in caption text for graphs because those '\' symbols would be written to the *∗.key* file and would ruin the performance of the *keytrans.exe* program.

F8≻[P]–> [P]layback a previously saved keystroke file

Several *∗.key* files have been supplied with your diskette[16], and you can create as many as you please. Use this [P] pick to replay them. The fastest way to learn to use this program is to replay a few of these *.key* files and study the way the program responds to various commands that an experienced user typed. Use the F8≻[d] option to search for names of *∗.key* files on your disk drive(s).

You will be asked to type the name (without the *.key* tag) of the file you wish to replay. You will then be prompted to hit the space bar to continue. From then on, the program will take its input from the *.key* file instead of from the keyboard. Each input character or line will be displayed as it is read from the file. Before responding to each command, the program waits for you to hit the space bar so that you can read the screen and learn what actions are performed in response to the commands.

After each command is read from the *.key* file, you will see some instructions in a box in the lower right of the screen. If you hit the space bar, the single command will be implemented. If you hit the 'S' key [17], playback mode will be aborted and the *.key* input file closed. It is possible to abort playback mode at *any* point, for example during a caption production operation. You must complete whatever process was in progress after you abort playback mode. The instructions on page 25 suggest keys to type if you cannot figure out what process you are supposed to complete.

---

[16]See chapter 6 on page 121 for a catalog of all such files.

[17]Note: upper case 'S'.

If you hit the ’>’ key, then the program will no longer wait for you to hit the space bar before executing each command. Instead, commands will be executed at maximum speed while most screen messages are suppressed. Use this feature to reproduce the results of a previous session quickly and automatically. Each of the many *.*key* files supplied on your diskette, played back in this rapid fashion, will produce an interesting graph. In this way you can accumulate a library of pictures, corresponding to problems discussed in this manual. These can be saved using the F1≻[s] option for subsequent instantaneous recall (*via* the F1≻[R] option).

If you pressed the ‘>‘ key to invoke rapid replay, and now wish to return to halt mode, in which you must hit the space bar to trigger execution of each single command, then press the ’<’ key. A screen message tells you this, but it may flicker too rapidly for you to read it. If you press the ‘<’ key while an integration or nullcline process in progress, the program will not return to halt mode until that process ends. However during replay, when an integration, finite difference iteration, nullcline location, or steady state point hunt is in progress, the ‘r’ key (to turn progress reports off and on), and the ‘q’ key (to abort the calculation) still function effectively.

5.11 Instructions for moving the graphics cursor

From either the F4 or the F5 menus, you can invoke a process to locate coordinates on the plane using a graphics cursor. *This makes sense only in the phase plane context*, but will act if selected regardless of that constraint. The cursor location software requires that the y_1 component is graphed on the horizontal axis and the y_2 component is graphed on the vertical axis. Here are the instructions printed on the screen when you are in “verbose menu” mode and invoke the graphics cursor.

Move the cursor using the arrow keys on the numeric pad.
The Home, End, Pg Up, and Pg Dn keys move diagonally.
Cursor appears when an arrow key is pressed.

The '+' and '−' keys increase and decrease
cursor move step sizes, respectively.

The SPACE key returns the present cursor location to
the calling program for use as initial conditions, arrowhead
locations, nullcline and steady state point searches, etc.

Hit the 't' key to have coordinates of cursor typed on screen.
Hit the 'k' key to type in coordinates of cursor.
The 'M' key toggles the mouse on and off
The 'm' key selects the mouse for cursor movement, if enabled

Hit the h key to recall these instructions.

These instructions are self-explanatory. When the cursor is alone on the screen, you have to remember to hit the space bar (or the enter key) to select the present cursor position and return to the calling procedure. Moreover, you have to remember to use the '+' and '−' keys to change the jump size of cursor movement. Holding an arrow key down produces continuous flow motion of the cursor. The instructions above are not printed if you are in terse menu mode. Whether you are in terse or verbose menu mode, the 'h' key will recall the instructions at any time.

↝ The "mouse options" mentioned above work only if you have a Mouse Systems Mouse (or one compatible with it) connected to the first serial port (com1:) and its port is activated[18] before the BIOGRAPH program is initiated. The three buttons on the mouse act as follows:

- Keep the right–most button depressed for fast coarse cursor movement.
- Press the middle button to accept the present cursor location and to return to the process that initiated cursor movement.

[18] To activate the first serial port, issue this MS-DOS command before starting the BIOGRAPH program:
MODE COM1:1200,N,8,1,P ↵.

- Press the left–most button to deactivate the mouse temporarily and to use the cursor arrow keys, to use the 't' and 'k' options, or to recall the cursor help instructions by hitting the 'h' key.

Note that cursor movements made using the mouse leave no useful trace in a $*$.*key* file you may be recording. If a $*$.*key* file is open while you are using the mouse, end each cursor movement process by pressing the left–most button, then hitting the 'k' key to type in the cursor's coordinates (which are displayed when you hit the 'k' key). These typed coordinates are written to the $*$.*key* file.

Chapter 6

Guide to the *.key playback files

The "player piano" feature of the BIOGRAPH program – the **.key* keystroke files it can create then replay – was described briefly on page 18 and in detail on pages 114—118. A major function of the **.key* playback files is to supply step-by-step demonstrations of BIOGRAPH actions. In addition, these files provide sources for important diagrams that are economical in storage requirements. In this chapter we catalog the **.key* files supplied on the diskette (with comments in some cases) to make it easier for you to take advantage of BIOGRAPH's player piano capabilities.

For increased speed, integrations are generally performed without exhibiting progress reports. The programs were "recorded", so that the integrations work properly. Nonetheless, progress reports can be requested at any time in the usual way, by toggling with the "r" key.

A few "mistakes" have been retained in the *.key* files, for it is instructive to see how one recovers from inevitable false starts.

Note that an earlier version of the BIOGRAPH program used little arrows to mark the nullclines. That version was used to produce the figures in MDP. In **.key* files that show how to redraw some of these figures, no attempt has been made to add arrowheads to the more esthetic line segments now used to show nullclines.

.pic files take a lot of disk storage space; *.key* files do not. Thus *glycosc.pic*, for example, could be erased, since *glycosc.key* will reproduce this picture.

List of .key files supplied

picshow Successively displays all the *.pic* files that are supplied on the diskette.

2d_camp (*select* = 2) Shows a limit cycle and trajectories approaching it for a quasi–steady approximation to our cAMP oscillator model.

camp3osc (*select* = 3) Produces Figure 1.1 on page 2 of the manual, which is essentially Figure 6.5 of the text. Presents the solution to exercise $\aleph 1_6$ in the next chapter[1].

liver (*select* = 4) Reconstructs the essence of Figure A5.12(b) on page 265 of MDP. Note how the flow arrows are first drawn for a rather large area of the phase plane, even including some area outside the first quadrant, to obtain an overview of the flow trajectories. Later attention is "zoomed in" to the area of greatest "action".

acubic (*select* = 5) Reproduces Figure 3.1 on page 37 of the manual and thus illustrates many aspects of phase plane plot construction.

giermein (*select* = 6) Generates the phase portrait of Figure 2.1 on page 17 of the manual.

glycosc (*select* = 9) Solves Exercise $\aleph 22(a)_6$ in the next chapter. It demonstrates how to deal with a situation where the solution varies very rapidly over part of the time domain. The general strategy is to integrate the rapidly varying portion separately and with high accuracy. A number of preliminary trial integrations gave the necessary understanding of the behavior of the solution. Remarks:

- For high accuracy *etol* and *hmin* are made very small.
- Suitable limits for the phase portrait were obtained from earlier trials with automatic scaling and F6≻[e].
- The large number of 1800 sample points was employed to ensure an accurate graph.

[1] This refers to exercise number 1 in this manual associated with chapter 6 in MDP.

- The first and second integrations were carried out for 8.65 and 348 time units because it had been previously ascertained that these were the durations of the fast and slow phases of the solution.
- After the limit cycle was drawn, the same data could have been used to make the window plots. A reintegration in one fell swoop was employed however, so that the automatic scaling feature could be utilized.

chaosc (*select* = 10) Produces figure 3.2 on page 49 of this manual in about 9 minutes of integration by a PC with an 8087 chip, showing tens of chaotically varying wave forms that result from the biochemical equations (3.53)—(3.57) on page 49 of the manual. If a few waveforms suffice to convince you of the chaotic nature of the solution, stop the integration early by striking the 'q' key.

When dealing with ODE systems that exhibit deterministic chaotic behavior, as this one does, the truncation error bound parameter, *etol*, must be set to a very small value (10^{-8}, say). This is because even the slightest errors in tracing trajectories will eventually be amplified into large errors that give a false prediction about the qualitative behavior of a solution. Smaller values of *etol* postpone, but do not eliminate this defect. Small values of *etol*, however, cause lengthy computations.

goldhart (*select* = 10) generates a two dimensional projection into the (β, γ) plane of a phase portrait for the Decroly–Goldbeter coupled feedback loop oscillator. The "strange" attracting set, projected into the (β, γ) plane appears as criss–crossing heart–shaped cycles.

bistab (*select* = 11) With the case *catrig.par*, reproduces the essence of Figure 9.12(c) on page 212 of MDP. Noteworthy is the use of backward integration to obtain the important unstable trajectories that emanate from the saddle point.

lorenz (*select* = 12) Once parameters and plot scaling conditions are set, the program integrates for several minutes to produce graphs of chaotically oscillating solutions of the deceptively simple Lorenz equations, (3.67)—(3.69), on page 54 of this manual.

owl (*select* = 12) re-graphs the solution computed by *lorenz.key* as a surprising two dimensional projection into the (y, z) plane of the three dimensional phase portrait. Use *owl.key* only after playing back *lorenz.key*.

lesson1 (*select* = 21) provides a simple introductory lesson, described on page 8, showing the steps necessary for solving Exercises $\aleph 1_2$ and $\aleph 2_2$ in the next chapter.

popgen (*select* = 25) provides the solution to $\aleph 1_3$ which involves the population genetics case described on 72

perdoub (*select* = 21) Produces Figure 4.1 on page 65 of the manual, and thus illustrates the use of the "superimposed epochs" feature (discussed in Section 4.1).

bifurc8 (*select* = 21) In a very lengthy calculation this file produces the classical "pitchfork bifurcation diagram" that represents in some detail the qualitative behavior of equation *(MDP:-2.7)* that was the principal subject of Chapter 2 in MDP. The "superimposed epochs" feature is used extensively here; one hundred epochs are graphed.

popgen (*select* = 25) Presents the solution of exercise $\aleph 1_3$ in the next chapter. Only when they are changed to accomplish parts (b)-(d) of the exercise is it completely verified that the parameters of *popgen.par* are appropriate to reproduce Figure 3.3 on page 43 of MDP.

insect (*select* = 26) Produces the complex phase portrait of Figure 4.2, page 73 of the manual.

arrows A sample "macro" playback file. Use it when constructing a phase portrait to generate arrowheads spaced at equal time increments of 0.5 units along a trajectory. To do so, position the cursor at the beginning of a trajectory you want to compute. Then use the F8≻[P] menu pick to playback the *arrows.key* file.

Chapter 7

COMPUTER EXERCISES

7.1 General remarks

For each chapter, the first exercises are generally intended to help you to master the use of the BIOGRAPH program, typically by asking you to reproduce the graphs or phase plane diagrams already given in the book. The fact that "answers" are given in full will help you gain confidence. These exercises also permit you to obtain various results that would otherwise have to be accepted on faith. Note that requests to reproduce a phase-plane diagram in the book are not meant to imply that every single trajectory must be reproduced; a few representative examples will suffice.

A major aim of computer simulation is to obtain intuition about the behavior of a model by examining the effects of parameters on properties of the solution. An example is determining the effect of flow rate on the time it takes a chemostat to approach steady state. We shall provide a number of such examples. In every case, we suggest that the following *PARAMETER INFLUENCE PROCEDURE* be followed. (i) Use your unaided intuition to guess the effect of the parameter; (ii) perform a number of computer simulations to see what the effect actually is; (iii) try to understand the reasons for the results you obtained, particularly if the results are not similar to those anticipated in (i).

Full treatment of a number of the more comprehensive exercises often amounts to a small project, and sometimes a fairly large one. In such cases the student may need to be content merely with making a serious beginning in the time available. Use your judgement.

Some questions are deliberately not precisely formulated; such vagueness is almost always present in "real" problems. You should then choose some reasonable precise formulation, and briefly defend the choice.

Some exercises request the reader to check others' simulation results concerning the behavior of solutions to a set of differential equations in a certain range of parameters. *You should not expect to duplicate this parameter range exactly*, because the numerical scheme used by BIOGRAPH is doubtless different from that used by the original authors. For example, Exercise $\aleph 29_6$ reports that Decroly and Goldbeter found "birhythmicity" for the parameter k in the range $(3.52, 4.04)$. It could well be that this range is say, $(3.51, 4.06)$ for the numerically integrated differential equations used in the BIOGRAPH program.

Many of the results for which verification is requested are quite new. Both human blunders and subtleties of numerical analysis could have given rise to incorrect conclusions. Like the repetition of a new experimental finding, the recalculations requested here actually advance the scientific enterprise a bit, by confirming others' assertions. This means, however, that perhaps the effort at "confirmation" will fail not because of an error by the student but because the results to be "confirmed" are in fact incorrect! No attempt has been made by the authors of this manual deliberately to set booby traps, but not every requested calculation has been carried out in full so that such traps may in fact exist. This gives the exercises a little of the flavor of true research.

As will be obvious from reading them, some exercises do not require use of the computer, but rather concern analytic matters that arise naturally during attempts to fully exploit the power of numerical simulations.

7.2 Numbering and reference conventions

"MDP" or "the text" is used to refer to the accompanying volume *MODELING DYNAMIC PHENOMENA IN MOLECULAR AND CELLULAR BIOLOGY*. Exercises in this manual are numbered consecutively by MDP–Chapter in the form $\aleph 1_2, \aleph 2_2, \ldots, \aleph 4_2$, then $\aleph 1_3, \aleph 2_3, \ldots, \aleph 4_3$, so that they can be distinguished from exercise numbering in MDP.[1] According to our numbering scheme, for example, exercise $\aleph 5_4$ refers to the fifth exercise in this chapter that is associated with chapter 4 of

[1] The decorative symbol ℵ is "aleph", the first letter of the Hebrew alphabet.

MDP.

This chapter comprises a number of parts. There are exercises corresponding respectively to most of the Chapters and Appendices of MDP, plus a number of miscellaneous exercises. There is also a section with partial answers and comments.

Equation numbers in this manual are written in the style (4.23), while the equation number *(MDP:4.3, p52)* refers to equation 3 in chapter 4 on page 52 in MDP.

We will refer to BIOGRAPH cases, e.g. the case "perdoub". Quotation marks should not be typed when instructing the computer to select this case.

References cited in the style of (Smith, 1983) refer to those provided at the end of MDP. References distinguished by an "asterisk", in the format (Jones, 1981)*, are provided at the end of this manual.

7.3 Exercises associated with specific chapters of MDP

No computer exercises are provided for chapter 1 of MDP which, from a mathematical standpoint, is dedicated to a review and application of basic ideas of calculus.

Exercises for chapter 2 of MDP

Exercise $\aleph 1_2$

Use case "perdoub" (*select* = 21) and four windows. For each window's vertical scale, make sure that $0 < x_n < 1.0$, and set the common horizontal scale to range from 0 to 30.

(a) Set $r = 0.5$, and $x_0 = 0.9$. Make the graph (which should appear in window 1). Make a second graph for $r = 0.995$ using a solid line. Compare with Figure 2.2 on page 21 of MDP.

(b) In window 2, make the graph for $r = 2.5$, $x_0 = 0.7$. Compare with Figure 2.5 on page 22 of MDP. Erase the graph and show on a new graph that the same steady state is approached in oscillatory fashion for two other initial conditions.

(c) Using window 3, graph the solution for $r = 3.25$, $x_0 = 0.9$. Compare with Figure 2.8, MDP page 26. Erase the graph, and try $r = 3.5$ with the same initial conditions but 100 generations. Try $r = 3.545$. Is your result in accord with Figure 2.10 of MDP?

(d) Again with 100 generations, graph the solution for $r = 3.75$, $x_0 = 0.4$ in window 4. Compare your result with Figure 2.9 of MDP. For practice, label the graphs with the parameter values employed.

Playing back the file called *lesson1.key* as described on page 8 will illustrate all the commands needed to do this exercise.

Exercise $\aleph 2_2$

Employ BIOGRAPH's calculational and graphical capabilities to help carry out the following exercises on pages 32–33 in MDP:

(a) Exercise 3

(b) Exercise 4

(c) Exercise 5

(d) Exercise 6

Exercise $\aleph 3_2$

Maynard Smith (1968), Chapter 2F, suggests equations for the numbers X_n and Y_n of prey and predator in the nth generation. Firstly, in the absence of predators, it is assumed that

$$X_{n+1} = RX_n \left(1 - \frac{R-1}{R} X_n\right).$$

(a) Show that this equation is equivalent to *(MDP:2.5, p17)* if the prey population is measured relative to its steady state value.

(b) In the presence of predators, the equations are taken to be

$$X_{n+1} = RX_n\left(1 - \frac{R-1}{R}X_n\right) - CX_nY_n, \qquad Y_{n+1} = rX_nY_n.$$

What has been assumed?

(c) Maynard-Smith (1968) shows that the populations appear to oscillate with a period of about a dozen years if $R = 1.5$, $C = 0.5$, $r = 2$. Note that with these parameters, if the prey is at its undisturbed steady state then in one generation half the prey population will be eliminated, and the predator population will double. In the absence of predators, the prey monotonically approaches a steady state. Initial conditions were taken to be $X_0 = 0.5$, $Y_0 = 0.2$. Use the *select* = 26 case[2] to graph the solutions in this case, for one or two hundred generations. Examine the effect of changing initial conditions.

(d) Try other parameter values and initial conditions, in an attempt to reveal different types of qualitative behavior. Are there always oscillations? What affects their period and amplitude? Can you understand why these effects occur? What effects might predators have on chaotic prey oscillations? — These are some questions that might be asked. What are other interesting questions?

(e) It is doubtful that all the interesting types of behavior will be found purely by computer experimentation. Thus perform a linear stability analysis of the two steady states, using the two-variable Taylor approximation *(MDP:A2.8, p234)*. To make the analysis a little easier, first make the change of variables $Z_n = CY_n$, which shows that C merely provides a scale for Y_n. Instead of using R and r as the two remaining parameters, use R and $\beta \equiv (R-1)/2r$. If you find the calculations difficult see Maynard Smith (1968) whose Figure 10 gives an almost complete picture of stability behavior in the $\beta - R$ plane. (The region $R < 1 + 2\beta$, for example, where there is no non-trivial equilibrium, is not shown in Figure 10.)

(f) Armed with the results of the stability analysis make a further investigation of qualitative behavior.

[2] Use the file *insect.par* as a template, but reset the values of all the parameters to those appropriate for this problem. The *select* = 26 case is described on page 73.

(g) It is not unreasonable to assume that the predator birthrate in a given year depends primarily on the predator's food consumption in the previous year. Hence, investigate the situation in which the predator equation of part (b) is replaced by

$$Y_{n+1} = rX_{n-1}Y_n.$$

As a first step, conjecture the effect of the change in the model on the oscillations demonstrated in part (c). Then use the standard parameters in case *diffquad.par* to see a "phase plane" graph of the appropriate solution. To check periodicity, replot the computer variable $y2$ in window 1 as a function of time, for the first 300 generations. Verify that indeed *diffquad.par* supplied the appropriate equations.

Exercise $\aleph 4_2$

(a) Plot representative results of Exercise $\aleph 3_2$ in the analog of the phase plane for differential equations. In particular, see Figure 4.3 on page 73 of this manual.

(b) Write a brief essay on some of the similarities and differences between "phase-plane" plots for differential and difference equations.

Exercise $\aleph 5_2$

The *select* = 21 cases "*perdoub*" and "*bifurc8*" on pages 68, 65, and 66 demonstrate the complexity of the solutions to the simple difference equation *(MDP:2.7, p17)*. Mathematicians are trying hard to understand this complexity, but it is not an easy task. For example, the well-known paper of Li and Yorke (1975)* proved that "period 3 implies chaos" in that if there is a solution of period 3 there is in fact an infinite number of periodic solutions[3] as well as an uncountably infinite number of asymptotically aperiodic solutions.

[3] Where N is any integer greater than 1, there is at least one solution having period N and no shorter period.

Show that nonetheless there can be a stable solution of period 3 by considering *(MDP:2.7)* with $r = 3.83$. How does this result fit in with the results of bifurc8? Note: For rigorous validation of the numerical evidence concerning the behavior of solutions to *(MDP:2.7)*, see for example Jakobson (1981)*.

Exercise $\aleph 6_2$

(a) Consider the equation

$$x_{n+1} = 1 - ax_n^2 \tag{7.1}$$

with $a = 1.9408$, and $x_0 = 0.3$. Show that there is a lengthy transient period after which the solution suddenly becomes periodic[4]. Calculate a few generations, and then employ F4≻[f] to set the initial conditions to the final value obtained in the previous integration. Repeat until periodicity appears. Show that the same "chaotic transients" are found for most other initial conditions.

(b) Show that a similar phenomenon appears for the two dimensional "Henon map"[5]

$$x_{n+1} = 1 + x_n - ay_n^2, \qquad y_{n+1} = bx_n, \tag{7.2}$$

with $b = 0.2$ and $a = 1.45$, where there is an attractor of period 5. Take initial conditions satisfying $|x_0| < 1.3$, $|y_0| < 1.3$ In addition to ordinary graphs of x_n and y_n *vs.* n, plot a "phase plane" for $|x_n| < 1.2$ and $|y_n| < 0.2$. In the latter case do not plot the first 10 iterations and stop plotting as soon as the solution becomes nearly periodic. See if you can discern an ordered pattern in the transient phase plane portrait.

See Kantz and Grassberger (1985)* for an analysis of the type of phenomena exemplified in this exercise.

---

[4]Use the *select* = 21 case, equation (4.3) on page 68 with file *perdoub.par* as a template, but reset the values of all the parameters to values appropriate for this problem.

[5]Use the *select* = 29 case, equation (4.22)—(4.24) on page 78 with file *diffquad.par* as a template, but reset the values of all the parameters to values appropriate for this problem.

Comments

Maynard Smith (1968) presents a model, ascribed to A.J. Nicholson and V.A. Bailey, for a host insect that is attacked by a parasitoid. In the case considered only one parasitoid survives in the host larva so that such a larva can give rise either to a single adult host or a single adult parasitoid. The following is a generalization of the Nicholson-Bailey equations, due to Hassel and Varley (1969)*.

$$X[n+1] = \kappa X[n] e^{-a|Y[n]|^{1-m}} \tag{7.3}$$

$$Y[n+1] = \kappa X[n] \left(1 - e^{-a|Y[n]|^{1-m}}\right) \tag{7.4}$$

Here $X_n \equiv X[n]$ denotes the number of hosts per unit area in generation n, and $Y_n \equiv Y[n]$ denotes the corresponding number of parasitoids.

Equations (7.3) and (7.4) are provided in case "bupalus" (*select* = 24, page 71). These equations will be the subject of the next four exercises.

Exercise $\aleph 7_2$

This exercise outlines a derivation of the host–parasitoid equations (7.3) and (7.4). To that end, let us fix our attention on a unit area. Let α be the probability that a given host in that area will be parasitized, during a given generation, by a single parasitoid.

(a) We shall employ $(1-\alpha)^{Y_n}$ as a formula for the probability that a given host is not parasitized. What has been assumed?

(b) Approximate the result of (a) to obtain the result that the probability that a given host is not parasitized is $\exp(-\alpha Y_n)$. What are you assuming?

(c) Suppose that there are equal numbers of male and female parasitoids and hosts. Let each female host lay $2E$ eggs that survive to become pupae (and therefore can serve as possible abodes for parasitoids). Assume that the same fraction F of hosts and parasitoids survive the winter until the next season. Finally accept that the empirical relationship found by Hassel and Varley (1969)*

$$\alpha = aY^{-m}$$

describes the interference between parasites (possibly due both to mechanical interference and to territoriality). Thereby derive (7.3) and (7.4), with $\kappa = EF$.

Comments

Maynard Smith (1968) depicts population data from a paper of G.C. Varley for the moth *Bupalus* in German pine forests during a 60 year period. There are thousand-fold irregular fluctuations in density with periods of 5-10 years. The populations can increase tenfold in a year, which suggests that κ is of the order of magnitude 10. This is the type of data that we wish to explain.

Exercise $\aleph 8_2$

This exercise concerns the case $m = 0$ (no interference between parasites) that was discussed by Maynard Smith (1968).

(a) Find the steady state points and examine their stability. Show in particular that divergent oscillations are predicted for small disturbances from the steady state in which both hosts and parasitoids are present.

(b) It is "well known" that if $m = 0$ then the full non-linear model predicts either extinction or unbounded growth. Check this assertion with some computer simulations.

Exercise $\aleph 9_2$

Consider the case of general m. (Realistic values of m range between 0 and 1.) Extend the analysis of the previous exercise. By considering some specific parameter values, compare your steady–state and stability results with those yielded by the computer (using F5≻[c]).

Exercise $\aleph 10_2$

Let $a = 0.1$ and $\kappa = 0.5$. Run a series of simulations[6] for 100 generations starting with $m = 1$ and decreasing m by steps of 0.1 to $m = 0.3$. Plot both host and parasitoid densities on the same graph, using different line styles.

(a) Write out your predictions concerning the results. Will there be oscillations? If so, for what parameter ranges? What will be the rough magnitude of the oscillations? Will peaks in host density lag or lead peaks in parasitoid density? Always? By how much? In answering, use the analytic results requested in the previous exercise, and/or rely on your raw intuition.

(b) After the simulations, compare the results with your predictions. "Post facto", what new insights can you offer concerning the host–parasitoid interaction? In particular, can you explain the result (pointed out by Hassel and Varley (1969)*) that relatively large m provides a stabilizing influence?

Exercise $\aleph 11_2$

(a) Use the parameter values of the case "bupalus". Change the plot so that host and parasitoid densities are displayed on the same graph, as functions of time (generation number), for a period of 100 years (which is about as long as one can reasonably expect to have data). "Integrate and make the plot". (The integration runs for many hundreds of years, but only the first hundred should initially be plotted). Note the character of the results; quasiperiodic outbursts. Using F1≻[m], make several more plots of succeeding periods of 100 or 200 years. Can you obtain any further insights as to the nature of the results?

(b) Read in the case "bupalus" again, "integrate and make the plot". A phase plane plot is produced. Note the remarkable result! It seems possible to draw the general conclusions that (i) considerable regularity can underlay what appears to be rather irregular

[6] Read in the Bupalus case file and set its parameters and plotting style to appropriate values.

observations, (ii) for difference equations, although important theorems no longer hold (e.g. "trajectories" can cross) still the phase plane is an important tool for revealing qualitative behavior.

(c) Manipulate parameters to see if you can find more complex behavior than that displayed in the case "bupalus".

Comments

For an introduction to the literature on this type of equation, see Jensen and Procaccia (1985)* and the references cited therein. It would be an excellent project for the more mathematically minded to survey this literature and to employ the computer to exemplify general theoretical results on the host–parasitoid problem. To generate enthusiasm we mention a few salient results. For certain parameter regimes as illustrated in Exercise $\aleph 11_2$, quasiperiodicity is manifested when the solution wanders irregularly on a smooth "limit cycle". In general, for other parameter regimes there is a recurrent visitation of points on the limit cycle and the solution is periodic. Chaotic solutions can occur when parameters are such that the smooth "limit cycle" develops wrinkles on all scales and becomes a so-called *fractal* structure.

Exercise $\aleph 12_2$

(a) Use case *insect.par* of *select* = 26 to investigate the behavior of

$$x_{n+1} = 1 + y_n - 1.4\, x_n^2, \qquad y_{n+1} = 0.3\, x_n.$$

Employ various plotting methods to reveal something of the seemingly random jumping around of the solution on a complicated set of points (which exemplify a "strange attractor").

(b) Replace "1.4" and "0.3" in part (a) by parameters a and b, find the steady states analytically as functions of a and b, and investigate their stability.

(c) Employ the analytic results of (b) as a starting point for further numerical investigation of the equations of (a). Your goal should be to discover different types of qualitative behavior.

For a report of an investigation of the equations of (a) by a father and son team, see Thompson and Thompson (1980)*.

Exercises for chapter 3 of MDP

Exercise $\aleph 1_3$

(a) Verify that as it stands case *popgen.par* (*select* = 25) will reproduce the dashed curve in Figure 3.3 *(MDP:p43)*, except that 40 generations are plotted instead of 20.

(b) On the same graph, reproduce the solid curve of Figure 3.3. What is the value of P_{40}?

(c) For the parameters of (b), graph the solution with the initial condition $P_0 = 0.1$.

(d) Erase the previous graph. Make a new graph of part (c), but for 80 generations.

(e) Put a detailed caption on the graph of (d) and label the axes.

Note: The solution to this exercise is worked out in the playback file *popgen.key*.

Exercise $\aleph 2_3$

(a) Reproduce Figure 3.4 on page 44 in MDP, and add graphs for $p_0 = 0.4, 0.74, 0.76, 0.9$.

(b) Verify that the same qualitative behavior is obtained for $K = -0.5$, $L = -1.0$.

Exercise $\aleph 3_3$

Verify *(MDP:3.24)* by direct calculation.

Exercise $\aleph 4_3$

(a) Let ν be the fraction per generation of a alelles that mutate to A, and let μ be the corresponding fraction for $A \to a$ mutation. One certainly expects that

$$\mu \ll 1, \qquad \nu \ll 1.$$

With this assumption, give intuitive justification that to include the main effects of mutation, in *(MDP:3.3, p37)* the numerator should be supplemented by

$$\nu(\Phi_{Aa} p_n q_n + \Phi_{aa} q_n^2) - \mu(\Phi_{Aa} p_n q_n + \Phi_{AA} p_n^2)$$

(b) Show that *(MDP:3.4)* implies that *(MDP:3.6, p38)* should be modified to

$$p_{n+1} = \frac{p_n(1 - p_n K)(1 - \mu) + \nu(1 - p_n)\left(1 - (1 - p_n)L\right)}{1 - K p_n^2 - L(1 - p_n)^2}.$$

See Exercises $\aleph 5_3$ for a derivation of the generalized equation.

Exercise $\aleph 5_3$

This exercise is concerned with a more careful justification of the intuitive result found in the previous exercise.

(a) Consider Table 3.1 on page 37 of MDP. Show that if the survivors of selection are then subject to mutation the number of AA genotypes becomes the following multiple of N:

$$\Phi_{AA} p_n^2 \left(1 - 2\mu(1 - \mu) - \mu^2\right) + 2\Phi_{Aa} p_n q_n (1 - \mu)\nu + \Phi_{aa} q_n^2 \nu^2.$$

(b) Derive corresponding expressions for the number of Aa and aa genotypes. Show that the total number of alleles is unchanged by mutation.

(c) Complete the required justification by neglecting double mutations as you set up the counterpart of *(MDP:3.3)*.

Exercise $\aleph 6_3$

This exercise concerns the magnitude of the steady state value $\overline{p}_\nu$ at which mutation at rate ν keeps an a-allele that would otherwise become extinct. In both cases, $\overline{p}_\nu \ll 1$ and the steady–state equation should be only approximately solved with this in mind.

(a) We make the assumption that in *(MDP:3.4, p37)* $L = 0$, $K > 0$. State why this case may be described as "selection against a recessive" and show that

$$\overline{p}_\nu \approx \sqrt{(\nu/K)}.$$

Hint. If p is small, in the cubic for p^3 the term Kp^3 should be negligible compared to $-Kp^2$.

(b) Now let us assume that $K = 0$, $L < 0$. State why this case may be termed "selection against a dominant" and show that

$$\overline{p}_\nu \approx \nu(1 - L^{-1}).$$

(c) Since $\nu \ll 1$ (typically $\nu = 10^{-6}$), $\nu^{1/2} \gg \nu$. Why should mutation be so much more effective at combating adverse selection in the cases of a recessive allele?

(c) By computer simulation of specific cases, show that it is likely that the steady states of (a) and (b) are stable, even to large disturbances.

Exercise $\aleph 7_3$

Mutation holds otherwise unfit alleles in the population at low levels until a time when they may become much more frequent owing to a change in conditions. The object of this exercise is to estimate the time it takes for the previously rare mutants to become noticeable. One important application of this type of study is to the appearance of resistant strains of micro-organisms subject to antibiotics, or of weeds subject to herbicides.

(a) Suppose that when $\nu = 10^{-6}$, $K = 10^{-3}$ a recessive allele is at the steady state value of Exercise $\aleph 6_3$. Assume that conditions change so that $K = -10^{-3}$. Use the computer to calculate how many generations it takes for p_n to exceed 0.9.

(b) Repeat (a) for the following shifts of K: 10^{-2} to -10^{-3}, 10^{-1} to -10^{-3}, 10^{-2} to -10^{-2}.

(c) Repeat parts (a) and (b) for the case of the steady state value of Exercise $\aleph 6b_3$. Use shifts of L from -10^{-3} to 10^3, etc, that parallel the shifts in K of parts (a) and (b).

(d) Organize the results that you have obtained into an illuminating table and/or graph. If you think it would be helpful, perform other calculations to fill out the picture.

Exercise $\aleph 8_3$

This exercise requests further, primarily analytic, study of the effect of mutation for the case of selection against a recessive (Exercise $\aleph 6(a)_3$).

(a) First consider the situation in the absence of mutation. Show that when $L = 0$, $K > 0$, the steady state equation has a double root $p = 0$ and another root $p = 1$. Examine the stability of these roots: in particular show that $p = 0$ is on the border between stability and instability. Conjecture qualitative behavior. Perform a few simulations to check your conjecture. Are the results biologically sensible?

(b) In the presence of mutation, the governing equation takes the form

$$p_{n+1} = (1-\mu)f(p_n) + \nu g(p_n); \quad 0 < \mu \ll 1, \quad 0 < \nu \ll 1.$$

Look (in the case $L = 0$, $K > 0$) for a steady state solution of the form $\overline{p} = 1 + \phi$ where $|\phi| \ll 1$. Using the Taylor approximation *(MDP:A2:7b,p234)* find a first approximation to ϕ, showing in particular that $\phi > 0$. Note that the derivative of f is given in *(MDP:3.15, p39)*.

(c) Look for a steady state solution of the form $\overline{p} = \psi$, $|\psi| \ll 1$. Show that there are two roots of this nature, one the positive root given in Exercise $\aleph 6_3$(a) and one a negative root. Note the split of the double root that was found in the case without mutation.

(d) Let $p_n = \overline{p} + p_n'$. Show that the linearized equation for p_n' is

$$p_{n+1}' = \left((1-\mu)\frac{df(p)}{dp} + \nu \frac{dg}{dp}(p) \right)_{p=\overline{p}} p_n'.$$

Examine the stability of the sole biologically meaningful steady state. The result should be in accord with the simulations of Exercise $\aleph 6d_3$.

Exercise $\aleph 9_3$

Repeat the previous exercise with suitable variation in detail for the case of selection against a dominant (see Exercise $\aleph 6(b)_3$).

Exercise $\aleph 10_3$

The stability studies of the previous two exercises led to the determination of easily solvable equations for p_n when p_n is small. Use these equations to redo Exercise $\aleph 7_3$ in a partially analytic manner—in the spirit of MDP's treatment of the example given on the bottom of page 44 of MDP.

Exercises for chapter 4 of MDP

Exercise $\aleph 1_4$

(a) Use case "acubic" (*select* = 5, page 37) to reconstruct Figure 4.1 on page 56 in MDP.

(b) Draw several other curves on the same set of axes for different values of k, to test the notion that k^{-1} is the time scale for $C(t)$.

Exercise $\aleph 2_4$

Use case "acubic" (*select* = 5, page 37) to construct graphs that will provide a check for your conjecture in Exercise 4.6(e) on page 71 in MDP.

The object of the next two exercises is to begin an examination of the validity of the quasi-steady state assumption. We shall consider a case that was simulated in Schauer and Heinrich's (1979)* study of this matter. These authors employed the parameters

$$k_1 = 1mM^{-1}\sec^{-1}, k_{-1} = 0.2\sec^{-1}, k_2 = 0.8\sec^{-1}, \text{so that } K_m = 1mM.$$

Exercise $\aleph 3_4$

In this exercise, we shall always consider the situation in which $S_0 = K_m = 1mM$. Thus the time of order

$$t_C \equiv \frac{1}{k} \equiv \frac{1}{k_1 S_0 + k_{-1} + k_2} \equiv \frac{1}{k_1(S_0 + K_m)} \tag{7.5}$$

for the complex to reach a quasi-steady state *(MDP:4.19, p55)* is always of magnitude 0.5 sec. Keep in mind the result of*(MDP:4.32, p61)* that significant change in substrate concentration takes place in a time of magnitude

$$t_S \equiv \frac{K_m + S_0}{k_2 E_0} \tag{7.6}$$

where here $t_S \approx 2.5/E_0$ sec, when E_0 is measured in mM.

(a) The *select* = 7 case "enz_sub" on page 41 treats the situation wherein $E_0 = 10^{-2}mM$. Thus $t_S = 250\,\text{sec}$. Integrate and make the plot. (Integration time has been set for 50 sec, enough to show some variation in the dependent variables.) Now make the parameter *select* = 8, and superimpose the quasi-steady–state approximation to the problem. (Don't forget to hit F7≻[i] to set the initial condition for C given by equation (3.49) on page 46.) You should find that the solution was calculated with good accuracy.

(b) Repeat (a) when $E_0 = 10^{-3} mM$. You will have to remember

- to integrate, and plot, for about ten times as long;
- to change the vertical scale for the complex concentration;
- to check that the initial conditions and initial and final time are appropriate.

You should find that use of the quasi–steady–state approximation gives faster and more accurate results than it did in (a).

(c) Now try $E_0 = 10^{-1} mM$ for 10 sec. You should find that the quasi-steady state assumption is not very accurate, *even though* $t_C \ll t_S$! Can you pinpoint the source of the inaccuracy? If not, see the discussion below exercise $\aleph 4_7$

Exercise $\aleph 4_4$

Explore the accuracy of the quasi–steady–state assumption along the lines of the previous exercise, only take $S_0 = 10K_m$. Can you reach any general conclusions. If you have time, also try $S_0 = 100K_m$ and $S_0 = 0.1K_m$ for a variety of values of E_0, and see if you can discern a rule that will describe your results concerning the accuracy of the quasi–steady–state assumption.

Comments

The simulations show that the quasi–steady–state approximation is sometimes an excellent one, and sometimes is not very accurate. The condition $t_C \ll t_S$ *(MDP:4.60, p61)*,

$$k_2 E_0 / k_1 \ll (K_m + S_0)^2, \tag{7.7}$$

while obviously necessary to ensure the validity of the approximation, does not seem sufficient. We must examine matters more closely.

By contemplation, or much more easily by examining the simulations, we are led to consider the fact that an integral part of the quasi–steady–state assumption is the initial condition $S(0) = S_0$ on the substrate. Implicit in this condition is the assumption that the substrate level does not noticeably decrease in the relatively short time during which the complex attains its quasi-steady state. To make this condition explicit, we use the fact that the initial rate of substrate decrease is approximately $k_1 E_0 S_0$ — by *(MDP:4:7b, p53)* assuming that $C(0) = 0$. We estimate the ***relative*** decrease in substrate during the time interval of duration t_C by multiplying $k_1 E_0 S_0$ by $t_C \equiv k^{-1}$ and dividing by S_0.[7] The reader should verify that the result is $E_0/(K_m + S_0)$. Thus another ***necessary condition for the validity of the quasi-steady state approximation is***

$$\frac{E_0}{K_m + S_0} \ll 1. \tag{7.8}$$

In fact, condition (7.8) is virtually the sole condition required for the validity of the quasi–steady–state assumption, as is indicated by the following exercise.

Exercise $\aleph 5_4$

Show that if condition (7.8) holds then condition (7.7) holds.

Comments

The parameters studied by Schauer and Heinrich (1979)* were somewhat arbitrarily chosen. It is not easy to measure all three relevant parameters, but here are several cases. The first two come from Dixon and Webb (1958)*, p.110, the third from Fersht (1985)*, p.69.

(i) Hydrolysis by trypsin of benzoyl-L-arginine ethyl ester:

$$k_1 = 4 \times 10^6 M^{-1}\,\text{sec}^{-1}, \quad k_{-1} = 25\,\text{sec}^{-1}, \quad k_2 = 15\,\text{sec}^{-1}.$$

(ii) Hydrolysis by chymotrypsin of acetyl-L-phenyl-alanine ethyl ester:

$$k_1 = 10^6 M^{-1}\,\text{sec}^{-1}, \quad k_{-1} = 90\,\text{sec}^{-1}, \quad k_2 = 10\,\text{sec}^{-1}.$$

(iii) Hydrolysis by chymotrypsin of acetyl-L-tryptophan p-nitrophenyl ester (neglecting a relatively fast intermediate step):

$$k_1 = 6 \times 10^7 M^{-1}\,\text{sec}^{-1}, \quad k_{-1} = 6 \times 10^4\,\text{sec}^{-1}, \quad k_2 = 65\,\text{sec}^{-1}.$$

---

[7] A substrate change of $\Delta S \equiv 0.1mM$ is insignificant if the starting substrate value is $S_0 \equiv 100mM$ and significant if this value is $S_0 \equiv 0.2mM$. The meaningful (and dimensionless) measure of change is $\Delta S/S_0$.

Two more cases follow. These are suspect, because the information concerning a single enzyme comes from two sources, and it is not at all certain that the same substrate was used and/or that the experimental conditions were the same. Nonetheless, the parameter values may at least provide correct magnitudes. The values of k_2 come from p.16 of Bernhard (1968)*. For (iv) the other two parameters come from Hammes and Schimmel (1970)*; for (v) from Rubinow (1975), p.56.

(iv) Ribonuclease:

$$k_1 = 10^7 M^{-1}\,\text{sec}^{-1}, \quad k_{-1} = 10^4\,\text{sec}^{-1}, \quad k_2 = 10^2\,\text{sec}^{-1}.$$

(v) Liver alcohol hydrogenase:

$$k_1 = 5 \times 10^5 M^{-1}\,\text{sec}^{-1}, \quad k_{-1} = 10^2\,\text{sec}^{-1}, \quad k_2 = 10^3\,\text{sec}^{-1}.$$

Exercise ℵ6$_4$

For the five cases listed above, make a table of values of t_C and t_S. Assume that $S_0 = K_m$ and that $E_0 = 10^{-2}K_m$ and $10^{-4}K_m$.

With these realistic parameter values note the extreme "stiffness" that will generally occur in the kinetic equations, since t_C is generally so much less than t_S.

Exercise ℵ7$_4$

(a) Consider case (iii) above. Set $S_0 = K_m$. Check that the quasi–steady–state assumption works well if $E_0 = 10^{-4}K_m$. (Note: K_m is approximately $10^{-3}M$, so that we wish to set $E_0 = 10^{-7}M$. Among other tasks, we must set the upper limit for the graph of the complex. This limit should be less than or equal to the value for E_0. But if you try to write this limit as "1e−7" you will get an error message stating that the upper limit must be greater than the lower limit 0. The reason is that these limits are restricted to six decimal digits, so that the given limit is interpreted as "zero". The solution: change units, say to mM.)

(b) Try $E_0 = K_m$ and show that the quasi–steady–state assumption is not a good one.

(c) Explore intermediate values of E_0 to check criterion (7.8) more carefully. Are there instances where the quasi–steady–state assumption is not a good one, yet the observed time for the complex to reach a quasi-steady state is small compared to the observed time for near-exhaustion of the substrate?

Exercise $\aleph 8_4$

Check for one or two more of the above cases that the quasi–steady–state assumption will not be justified if (7.8) does not hold. You only have to integrate for a time of magnitude t_C, demonstrating that there is a significant decrease in substrate level.

Comments

As was mentioned in Appendix 7 of MDP (p. 279), the use of dimensionless variables is advantageous if extensive numerical studies are to be made, because the number of required parameters is reduced. All nondimensionalizations are equally effective in reducing the number of parameters. BIOGRAPH provides two different nondimensional versions of the enzyme-substrate problem under discussion and of the quasi–steady–state approximation to it. The second of these, denoted by IFDIM=2 (see page 42), is more useful. Here the parameter combination t_S of (7.6) is used as the time scale in defining the dimensionless time. Thus the substrate is significantly depleted in a time of magnitude unity under all circumstances. Moreover, we shall see that the dimensionless dependent variables are also of magnitude unity for all parameter values. This particular nondimensionalization is thus very convenient for our computer explorations, because once and for all we can set the scales for both the dependent and independent variables.

The dimensionless equations associated with IFDIM=1 employ t_C of (7.5) as a time scale. The IFDIM=1 equations are thus especially suited for examination of the solution during the period in which the complex changes relatively rapidly before settling into its quasi-steady state condition. A special dimensionless product concentration is defined for IFDIM=1, since very little of the total product is produced during this early period.

Let us consider the most frequently studied situation wherein $C_0 = P_0 = 0$. There are then five dimensional parameters: k_1, k_{-1}, k_2, S_0, and E_0. There are only three dimensionless parameters, as defined in equations (3.23)—(3.32): η, $\tilde{S}_0$, and $\tilde{E}_0$. In dimensionless terms, condition (7.8) for the validity of the quasi–steady–state assumption is simply

$$\tilde{E}_0 \ll 1 + \tilde{S}_0. \tag{7.9}$$

Exercise $\aleph 9_4$

(a) For $E_0 = 10^{-6}M$, $S_0 = 10^{-2}M$, calculate the dimensionless parameters of equation (3.29) for cases (iii) and (iv) from the list of rate constants preceding Exercise $\aleph 6_4$. Verify that the BIOGRAPH program computes these parameters correctly when the F7≻[i] option is used as directed on page 43.

(b) Check condition (7.9) by calculating two or three sets of parameter values for case (ii).

(c) Repeat (b) for case (iv) in the list preceding exercise $\aleph 6_4$.

(d) Sketch the region $\tilde{E}_0 \leq 0.01\,(1+\tilde{S}_0)$ in a graph of the $\tilde{E}_0$-$\tilde{S}_0$ plane. Also sketch this region in the $\log \tilde{E}_0 - \log \tilde{S}_0$ plane (which is often more convenient to use, since one is interested in parameter values that range over several orders of magnitude).

Exercise $\aleph 10_4$

Two different nondimensionalizations are detailed in the description of *select* = 7 (pages 41—42). As was explained above, the dimensionless variables were defined so that they would vary over an interval of magnitude unity. We have already explained the choice of scales t_C and t_S for the two different dimensionless independent variables. Try to provide rationalizations for the choice of the scales for the dependent variables. Compare your thoughts with the "answer" given on page 208 of this manual.

Exercise $\aleph 11_4$

It is good applied mathematical practice to check for the consistency of any approximations that one makes. Here we have essentially deleted the term dC/dt in the complex equation (3.21), and we hope that this simplification is justified except for a relatively short initial period. This exercise guides the reader through a consistency check. For a discussion of such checks and possible pitfalls accompanying them, see section 6.1 of Lin and Segel (1974).

Let us work with the $IFDIM = 2$ dimensionless equations, (3.46)—(3.48).

(a) Verify the quasi–steady–state assumption given in (3.41)

$$c = \frac{(\tilde{S}_0 + 1)s}{\tilde{S}_0 s + 1}. \tag{7.10}$$

(b) To check the consistency of this assumption, we write

$$c = \frac{(\tilde{S}_0 + 1)s}{\tilde{S}_0 s + 1}(1 + \Delta). \tag{7.11}$$

We suppose that our assumption is justified, and therefore that $\Delta(t)$ and its derivatives are small compared to unity. Given this, show that

$$\Delta \approx -\frac{\tilde{E}_0}{(\eta + 1)(\tilde{S}_0 + 1)(\tilde{S}_0 s + 1)^2 s}\frac{ds}{dT}. \tag{7.12}$$

(c) Thereby deduce

$$\Delta \approx \frac{\tilde{E}_0}{(\eta + 1)(\tilde{S}_0 s + 1)^3}. \tag{7.13}$$

What is the biochemical interpretation of the fact that $\Delta > 0$? Is your interpretation intuitively reasonable?

(d) Consider situations in which not much substrate has been consumed, so that $s \approx 1$. Show that if (7.8) holds then Δ is necessarily small.

This is the consistency we seek: if we assume that Δ and its derivatives are small then we obtain an expression according to which Δ is indeed small.

(e) Show that inconsistency can result if the reaction progresses far enough so that s becomes small. Such an inconsistency is in fact responsible for the poor accuracy observed by those who completed exercise $\aleph 3_4$. The existence of difficulty for large $\tilde{E}_0$ was indeed pointed out by Schauer and Heinrich (1979)*, but no clear biophysical explanation was supplied by these authors.

(f) We thus conclude once again, from a different point of view, that the quasi-steady state assumption is a good one provided that (7.9) holds. It is true that even if (7.9) holds the approximation may become a poor one if s has decreased so much that $s\tilde{S}_0$ is small compared to unity — if $\tilde{E}_0/(\eta+1)$ is of unit magnitude or greater. Show that such an eventuality can occur only when S decreases to magnitude K_m from an initial value that is far in excess of K_m. Thus this "exception" to the criterion (7.9) is generally of little practical importance. In any case criterion (7.9) indicates that there will be trouble if E_0 and S are both of magnitude K_m.

Exercise $\aleph 12_4$

The purpose of this exercise is to examine some of the effects of changing the integration parameters. Use case *enz_sub.par* with parameter values $E_0 = 0.01,\ k_1 = 10,\ k_{-1} = 0.1,\ k_2 = 20$. Plot $C(t)$, $S(t)$, and $P(t)$ for $0 \le t \le 10$, with scales $(0, 5\times 10^{-4})$, $(0, 0.1)$ and $(0, 0.03)$ respectively. Take $C(0) = 0,\ S(0) = 0.1,\ P(0) = 0$.

(a) Experiment with different *fixed* step sizes by taking hmin=hmax. Keep in mind the remarks on page 104 that the step size will be taken smaller than the common value of *hmax* and *hmin* unless you specify, via F3≻[#], that a sufficiently small number of sampled values is to be stored. For example, if you wish a fixed step size of 0.1 in the problem under consideration, you must set *hmin* = *hmax* = 0.1 and also specify that 100 values (or fewer) are to characterize the solution for $0 \le t \le 10$.

(b) Observe the graphs of the solution as *etol* is decreased from *etol* = 1 by steps of 0.1 until smooth curves are obtained for all variables. Try to explain as much as possible of the qualitative behavior of the "solutions" that you observe.

(c) Try to construct situations where changes in other integration parameters have a marked effect on the solution or on the time taken to obtain it.

Exercises for chapter 5 of MDP

Exercise $\aleph 1_5$

(a) Use the *select* = 1 *chemo.par* case with parameters suitable to Figure 5.8(b) on page 93 in MDP. Set the graph height to 4.0, and the width to 5. Put the cursor in the middle of the screen and locate both nullclines. Find the steady state point and check by hand that the data given is correct. Fill the screen with a number of flow arrows; compare with Figure 5.8(b).

(b) Reproduce the nullclines and at least some of the trajectories (including a few flow arrows) of Figure 5.8(a) in MDP. To do this, use F5≻[t] together with F5≻[a].

Exercise $\aleph 2_5$

Replot one of the trajectories from $\aleph 1_5$(b). Superimpose standard plots of $n(t)$ and $c(t)$ *vs.* t in a window on the phase portrait.

Exercise $\aleph 3_5$

(a) Examine the dependence on the flow rate q of the time it takes the chemostat to approach steady state. You will have to adopt some reasonable operational definition of "approach steady state". Moreover, you will have to keep initial conditions constant, for example with $C_0 = 0$ and N_0 some reasonable value.

(b) How do your results depend on N_0?

(c) Compare your results with the prediction of Exercise 5.9 on page 95 in MDP.

Exercise $\aleph 4_5$

Use the *select* = 1 *chemo.par* case to:

(a) Check the conjecture of Exercise 5.11(d) on page 96 in MDP by constructing a phase plane diagram. (See section 7.4 for a hint.)

(b) Repeat (a) for Exercise 5.11(e).

Exercise $\aleph 5_5$

In this exercise we re-interpret the chemostat equations *(MDP:5.9, p80)*. Take $C(t)$ to be the concentration not of a chemical nutrient, but of an organism eaten by a predatory species in suspension in the chemostat. That is, $N(t)$ = the total number of predators in the chemostat, while $C(t)$ = the number of prey organisms per unit volume in the chemostat.

Assume that the $C(t)$ prey species reproduces exponentially, supported by a chemical nutrient that is so abundant that it does not limit their reproduction. Thus, the effect of C reproducing can be modeled by adding the term $\rho C(t)$ to the right hand side of *(MDP:5.9b)* to obtain

$$\frac{dC}{dt}(t) = q\left[C_i - C(t)\right] + \rho C(t) - \frac{N(t)MC(t)}{y\left[K + C(t)\right]V} \tag{7.14}$$

(a) Using the same scaling as in MDP, reproduce the derivation of the dimensionless version of (7.14) to show that *(MDP:5.21b, p84)* should be replaced by

$$\frac{dc}{dt} = \xi - \gamma c - \nu\frac{\mu n c}{1+c} \tag{7.15}$$

in which $\gamma = 1 - \rho/(qK)$. This means that, as the reproduction rate of the "nutrient" bacteria increases, γ decreases. With this interpretation, negative values of γ make biological sense. The *select* = 1 case on page 30 includes equation (7.15) and the original equation for dn/dt.

(b) Show that the unique steady–state solution in which c and n coexist is

$$\overline{c} = \frac{1}{\mu - 1}, \qquad \overline{n} = \frac{1}{\nu}\left(\xi - \frac{\gamma}{\mu - 1}\right) \tag{7.16}$$

(c) Reproduce the linear stability analysis in MDP on page 85 and show that when

$$\gamma = \gamma_0 \equiv -\xi(\mu - 1)^2 \tag{7.17}$$

the value of β corresponding to *(MDP:5.33, p86)* is 0. In the three–dimensional (μ, ν, γ) parameter space, equation (7.17) specifies a two dimensional surface. For parameter values on that surface, linear stability analysis of (3.1) and (3.2) predicts that $\beta = 0$ in Figure A5.9 on MDP page 258 – that is, predicts the steady–state point is a linear center.

(d) For each of a few values of μ, use the BIOGRAPH program to construct some phase portraits for (3.1) and (3.2) when γ is close to γ_0. For values of γ either slightly greater than γ_0 or slightly less than γ_0, but not both, you should find limit cycles in the phase portrait. By numerical experimentation, try to characterize the values of γ that correspond to population oscillations, and, by using the F5≻[c] menu option, find out what kind of steady–state point $(\overline{c}, \overline{n})$ is for oscillatory behavior.

(e) Interpret the biological significance of your mathematical discoveries in part (d). That is, for what values of the "nutrient" bacterial reproduction rate, ρ, will population oscillations occur in the chemostat. Describe verbally what drives such oscillations.

(f) If you do part (d) correctly, you will see a so–called *Hopf bifurcation* occur. The Hopf bifurcation theorem specifies the general conditions required for this kind of bifurcation to occur and includes a recipe for theoretically determining parameter values for which *stable* limit cycles arise. See Odell (1980)* for a description of the use of this powerful theorem, and try to carry out the calculation described therein for determining when bifurcation to stable limit cycles occurs. The results should confirm your numerical experiments in part (d).

Exercises for chapter 6 of MDP

Exercise $\aleph 1_6$

(a) Use case *camp3osc.par* to reproduce Figure 6.5 on page 107 of MDP[8]. Note that the program operates with dimensionless units, so that 1000 seconds is 100 dimensionless time units according to *(MDP:6.2, p103)*.

(b) Note that the step size in the integration is frequently increased and decreased. Make a change to improve integration performance.

(c) Show how a close-up look can be taken of a portion of your graphs.

(d) You have obtained three graphs plotted in one window. Now plot each of the graphs in a separate window, using automatic scaling.

Exercise $\aleph 2_6$

Perform the calculations necessary to obtain the results of Figure 6.7 on page 109 of MDP for $\gamma_i/\bar{\gamma} = 1.5, 3, 40$. (Parameter values are those of Figure 6.6.)

Exercise $\aleph 3_6$

(a) Figure 6.8 on page 110 of MDP is schematic. Make a start at rendering Figure 6.8 quantitatively accurate (like Figure 6.15) by varying σ, keeping all other parameters fixed at the values used in Figure 6.5, and determining the extent of the oscillatory zone. Describe how the period of the oscillations and the shapes of the graphs change as σ is varied.

(b) Repeat (a) for σ fixed and k varying.

[8] The file *camp3osc.key* produces a solution to this exercise.

(c) Delineate the relay zone in Figure 6.8. Regard "relay" as occurring if an instantaneous elevation of extracellular cAMP γ produces a pulse wherein γ exceeds its initial value.

See section 7.4 for a partial answer.

Comments

The top photograph of the Frontispiece in MDP shows a number of aggregation territories. Here is one theory that partially explains how these territories arise.

In a certain region, some cell, C_1, will be most advanced in its development and hence will be the first whose parameters move along the developmental path into the "oscillation" domain of Figure 6.9 (page 110, MDP) or Figure 6.15. If it were isolated, a second cell C_2 might next start to secrete cAMP in an oscillatory fashion. But suppose that the frequency of pulses generated by C_1 is higher than that of C_2. Then C_2 is likely to receive a superthreshold stimulus of extracellular cAMP, and to secrete in response to this stimulus, rather than to carry out its own independent secretion program. (For example, consider a time interval during which the state of C_2 is represented by a point moving slowly up the vertical part of the limit cycle trajectory, on the left side of Figure 6.12, page 116 in MDP. A nearby center could then emit a cAMP pulse that would prematurely trigger C_2 to secrete.)

Thus "delayed" centers could be "taken over" by the earliest center, so that the delayed centers in effect act like ordinary relay cells. The earliest center C_1 represses possible competitors for a distance over which, roughly, C_1's cAMP secretion is large enough to exceed the signaling threshold of Figure 6.7 (p109, MDP).

The above explanation contains two implicit assumptions. It is the purpose of the next exercise to check these assumptions.

Exercise $\aleph 4_6$

(a) Employing the system *(MDP:6.3, p103)*, show that an oscillating cell can act as a relay cell. Do this by temporarily stopping the integration and then providing a superthreshold signal of extracellular cAMP, at an appropriate time during the normal oscillatory cycle.

(b) Investigate the question of whether the oscillatory frequency of a cell increases as its parameters move through the oscillatory domain along the developmental path of Figure 6.9 (p110 in MDP). Note: If you have done Exercises $\aleph 3_6$ (a) and (b), what is requested here is essentially implicit in the results you have already obtained.

(c) Carry out (a) and (b), but employ *(MDP:6.28, p124)* instead of *(MDP:6.1a, p102)*.

Exercise $\aleph 5_6$

Observations of Robertson and Drage (1975)* indicate that relay can be induced by constant signals as well as by pulsatile signals. The purpose of this exercise is to try to obtain such behavior by adding a constant source term S to the **right** side of *(MDP:6.1c)*, and in particular to show that there is a sharp threshold (Goldbeter, 1980, p.275). Use the *select* $= 3$ case "camp3osc", $s \equiv SK_s/K_pV \neq 0$.

(a) The parameters should be set for values that give relay with $s = 0$, and the initial conditions should be the corresponding steady state. Why? The horizontal scale should be set to about 10 time units. Why? (See page 212 for answers.) Show that relay occurs when $s = 2$ but does not occur when $s = 0.1$.

(b) Find the threshold value of s to within 0.1. Try to supply an intuitive explanation for the magnitude of the threshold, based on earlier findings.

Exercise $\aleph 6_6$

It has been observed that a sufficiently large cAMP stimulus does not produce any amplification (Grutsch and Robertson, 1978)*. Show this by simulation. Explain the phenomenon by means of the phase plane. (The explanation appears in Goldbeter 1980, p. 273.)

Exercise $\aleph 7_6$

Use the *select* = 2 "2d_camp" to reproduce the following figures in MDP

(a) Figure 6.12 (MDP, p116). Use F5≻[t] and F5≻[a] to place flow arrows on the limit cycle at equal time intervals.

(b) Figure 6.13 (MDP, p118). For Figure 6.13(b), two or three representative trajectories will be sufficient.

(c) Figure 6.14 (MDP, p121). Note that exactly the same phenomenon is depicted from different points of view in Figures 6.5 and 6.12 and in Figures 6.6 and 6.13.

Exercise $\aleph 8_6$

Repeat Exercise $\aleph 4_{A1}$, using the graphs of Figure 6.5 as a test case.

Exercise $\aleph 9_6$

Check Figure 6.15 (MDP, p125) by selecting parameter values in each of the various regions and obtaining graphs of the form of Figures 6.5 and 6.6. In particular, try to demonstrate the existence of two steady states in the situation of region E. (See Goldbeter 1980, p. 281.)

Exercise $\aleph 10_6$

Investigate the phase plane for the situation of region E in Figure 6.15 of MDP.

Exercise $\aleph 11_6$

This exercise uses the possibility of non-zero values of the parameter s in such cases as "2d_camp" and "3d_camp" ($select = 2,\ 3$) to model the experiments of Devreotes and Steck (1979) that are mentioned on p. 127 of MDP. The basic idea of the experiments is that it is possible artificially to add extracellular cAMP at such a constant rate as to "clamp" the extracellular cAMP concentration at any desired value. That is, the effect of phosphodiesterase is experimentally nullified by a sufficiently high injection rate of extracellular cAMP. By elegant radioactive labeling techniques, it is possible to distinguish the cAMP added by the experimenter from the cAMP that was produced and secreted by the cells.

Simulate the Devreotes and Steck (1979) experiments in the following way. First obtain relay by "turning on" a slightly superthreshold source as in exercise $\aleph 5_6$. Around the time that the response would usually start to decrease, simulate a second elevation in the extracellular cAMP concentration by increasing the strength of the cAMP source. It is anticipated that no significant delay will be found in the eventual decay of the response. But a marked delay was found experimentally. (Modifications to the model have been made that allow it to explain these experiments. See, for example, the theory of Martiel and Goldbeter (1984)*, based on desensitization of the cAMP receptor. Also see the ideas of Segel, Goldbeter, Devreotes, and Knox (1985, 1986)*.

Exercise $\aleph 12_6$

Experiments show (e.g. Gerisch and Wick, 1975) that the period of the autonomous cAMP oscillation markedly decreases with time.

(a) Can these results be explained by the "developmental path"? (Compare Exercise $\aleph 3_6$.)

(b) Are there other parameter changes aside from those in σ and k that might be used to explain the experiments?

(c) The cAMP oscillations can be phase shifted by extracellular cAMP pulses (Gerisch and Wick, 1975), *i.e.* the time that the maximum occurs (for example) can be advanced and retarded. Study the behavior of the model with respect to this phenomenon.

Exercise $\aleph 13_6$

There is very little detailed justification given for the equations *(MDP:6.1-3, p102)*. Provide such a justification. Assume that the intracellular conversion of ATP into cAMP takes place in a narrow "reaction layer" of width d (see Figure 7.1). Let the input of ATP into the reaction layer be at the constant rate F per unit time per unit area. Let E be the number of cyclase molecules per unit area of membrane.

(a) Consider the change during a short time interval Δt in the total number of ATP molecules in the reaction layer, using *(MDP:4.60, p70)*. Thereby derive *(MDP:6.1a, p102)* showing in particular that V is proportional to F and W is proportional to E.

(b) In deriving an equation for dB/dt justify and use the assumption that in a short time interval of length Δt, every molecule that is within a distance $\delta\Delta t$ of the membrane has a probability θ of changing from intracellular to extracellular or the reverse.

(c) Proceed similarly to derive an equation for dC/dt.

See Section 7.4 for a solution to exercise $\aleph 13_6$.

Comments

Chapter 4 of MDP discussed in some detail the quasi-steady state assumption for an enzyme substrate reaction. The discussion led to the classical way of obtaining the maximum reaction velocity and the Michaelis constant by using the Lineweaver-Burk plot. Other important roles for the quasi-steady state assumption have been demonstrated in Chapter 6. (i) A quasi-steady state assumption for the Monod-Wyman-Changeux allosteric dimer yielded equation *(MDP:4.60, p70)* for the rate of product formation. This was the basis for the key expression *(MDP:6.1d, p102)* in our theory for cAMP signaling. (ii) A quasi-steady state assumption allowed us to pass from the third order system *(MDP:6.3, p103)* to the second order system *(MDP:6.15, p111)*. Powerful tools of phase plane analysis were applied to *(MDP:6.15)*; these generated a much deeper understanding of the behavior of our model for cAMP signaling.

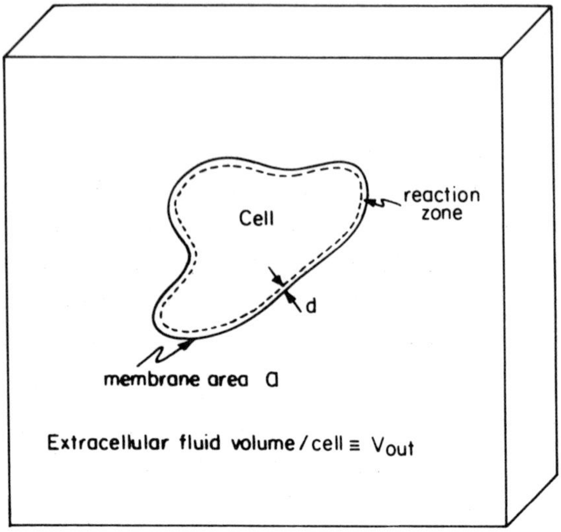

Figure 7.1: The reaction layer within a cell in which the conversion of ATP to cAMP takes place

It is natural to ask: When is a quasi-steady state assumption valid, and for what variables? These are usually difficult questions; no satisfactory general "method" is available to answer them. The next eight exercises will illustrate how such questions can be approached by a combination of analytic and computer methods. (Also see the computer exercises for Chapter 4.) Working through the bulk of these exercises would make an excellent project.

Note that our approach on pages 57–61 of Chapter 4 of MDP, culminating in condition *(MDP:4.33)*, cannot easily be extended to establish a corresponding condition that is necessary for the validity of the quasi-steady state assumptions used in Chapter 6. The major reason is this. If we set S =constant in *(MDP:4.13b)* we obtain an equation for C that we can solve easily and thereby find the time scale for complex change. The same device will not work for the far more complex equations of *(MDP:6.3)*. On the other hand, the cAMP results are of a qualitative character, while quantitative measurements of reaction constants often are the motivation for the quasi–steady–state assumption in the simplest enzyme–substrate reaction. Thus although the former case is more complex, there is the compensation that looser criteria can be used in defining a satisfactory approximation.

As the next exercise illustrates, one way around the difficulty is to use the computer in a "brute force" approach to the discovery of possible steady state approximations. For a given set of parameters, one can simply try such approximations and see whether or not they work.

Exercise $\aleph 14_6$

(a) Calculate the equivalent of Figure 6.5 (MDP, p107) when quasi-steady state approximations are made respectively on *(MDP:-6.3a)*, *(MDP:6.3b)*, and *(MDP:6.3c)*. You should find that assuming that α is in a quasi-steady state does not give a good approximation, but *either* β *or* γ can be assumed to be in a quasi-steady state. The *select* = 13 case is useful here. Try to solve the equations analytically if you have had a first course in differential equations.

(b) If quasi–steady–state assumptions are made on two of the three equations, the problem reduces to a single first–order equation. Try to demonstrate that such first–order equations can *never* have oscillatory solutions.

Exercise $\aleph 15_6$

Repeat part (a) of Exercise $\aleph 14_6$ when the relay of Figure 6.6 (MDP, p108) is examined instead of the oscillations of Figure 6.5. Use the *select* = 13 case.

Exercise $\aleph 16_6$

Show that the qualitative results of the two previous exercises remain true if the parameters are altered somewhat from those of Figures 6.5 and 6.6 of MDP. Attempt to gain some intuition concerning the role of parameters in rendering the quasi-steady state assumptions respectively more and less legitimate. Use the *select* = 13 case.

Comments

Our computer simulations have indicated that for certain parameters either β or γ can be assumed to be in a quasi-steady state. With this information to encourage us, let us attempt to find analytic justification for these quasi-steady state assumptions. In doing so, we shall examine the equations in their original dimensional form *(MDP:6.1, p102)*.

When indeed will the extracellular cAMP concentration C be in a quasi steady state? From Chapter 4 of the text we know that for C to be in a quasi-steady state its intrinsic rate of change must be rapid compared to that of both the intracellular cAMP concentration B and of the intracellular ATP concentration A. We must keep in mind that our discussion concerns the *model (MDP:6.1)*: it is the model that we are attempting to approximate. For the time being we do not concern ourselves with how cAMP changes in a real cell, although comparison with reality is of course the eventual goal of our theoretical work.

Extracellular cAMP changes through the agency of two processes, secretion and hydrolysis by phosphodiesterase. If only the latter process were present, its time scale would be easy to estimate. From $dC/dt = -k^*C$ (compare Equation *(MDP:6.1c)*) it follows as on pp. 55–57 of MDP that the time scale is $(k^*)^{-1}$. On pp. 55–57 we gave two ways to obtain this result. Here is a third way, which will shortly prove useful.

The characterization of the time scale in *(MDP:4.21, p56)* can be regarded as giving the time that it takes the function to undergo its maximum change, if the change takes place at its maximum rate. Another handy characterization of the time scale, employing the same general spirit, is

$$\text{Time scale of } f(t) \approx \frac{|\text{typical change in } f|}{|\text{typical value of } df/dt|}. \tag{7.18}$$

If f changes appreciably then

$$|\text{typical change in } f| =$$

$$|\text{typical largish value of } f - \text{typical smallish value of } f|$$

$$\approx |\text{typical largish value of } f|.$$

Thus

$$\text{Time scale of } f(t) \approx \frac{|\text{typical largish value of } f|}{|\text{typical value of } df/dt|},$$

or even, most conveniently, albeit somewhat less accurately,

$$\text{Time scale of } f(t) \approx \text{typical value of } \left|\frac{f}{df/dt}\right|. \tag{7.19}$$

In the case just considered, $dC/dt = -k^*C$. Here the quotient of C and dC/dt is always

$$\left|\frac{C}{dC/dt}\right| = \frac{1}{k^*},$$

leading once again to our familiar result.

Returning from our brief methodological excursion, we recall again that the hydrolysis of phosphodiesterase has the time scale $(k^*)^{-1}$. Turning to secretion we focus on equation *(MDP:6.1b)*. If only secretion were present then $dB/dt = -k_t^*B$, yielding a time scale of $(k_t^*)^{-1}$. Typical values of the two time scales are 2.5 sec, as in the caption of Figure 6.5 (MDP, p107). This figure shows how short 2.5 seconds is compared to the time scales for changes in α and β. Thus it is the relatively rapid nature of secretion and hydrolysis that suggests that a quasi-steady state assumption can be made on the extracellular cAMP concentration γ.

Exercise $\aleph 17_6$

Estimate how either k^* or k_t^* must be changed so that a quasi-steady state assumption is no longer valid for γ. Verify your estimate with a computer simulation.

Comments

The time scale for one of the two processes that affect the extracellular cAMP concentration C was easily estimable from equation *(MDP:-6.1c)* for dC/dt, while the other process could be estimated from its appearance in the equation for dB/dt[9]. No such easy methods seem apparent for estimating the time scales of A and B. Progress can be made, however, as we now show.

Let us focus our attention on relay. From Figure 6.6 (MDP p108) we see that in the first stage of relay, the dimensionless substrate concentration α decreases, while the dimensionless intracellular cAMP concentration β increases. Referring to the dimensional equations *(MDP:6.1a)* and *(MDP:6.1b)*, we observe that the cAMP synthesis term $W\Phi$ must therefore outweigh the other terms V and $-k_t^* B$ respectively. That is, in the first stage of relay, at least to a rough approximation,

$$dA/dt \approx -W\Phi(A,C), \quad dB/dt \approx W\Phi(A,C). \tag{7.20}$$

From equation (7.19), it follows that

$$\frac{\textit{Time scale of } A(t)}{\text{Time scale of } B(t)} \approx \text{typical value of } \left|\frac{A}{dA/dt}\,\frac{dB/dt}{B}\right|.$$

But from (7.20) and *(MDP:6.2, 6.4)*

$$\left|\frac{A}{dA/dt}\,\frac{dB/dt}{B}\right| \approx \left|\frac{A}{B}\right| = \left|\frac{K_s\alpha}{K_p\beta}\right| = \left|q\frac{\alpha}{\beta}\right|.$$

We note from Figure 6.6 (MDP p108) that typical values of α and β are the same, about 100. From the caption to Figure 6.5 we observe that typically $q \approx 100$. It follows that in the first stage of relay, because $q \gg 1$, the time scale for changes in β (or B) is small compared to the time scale for changes in α (or A). This indicates that a quasi-steady state assumption may be made on β.

Now consider the second half of the course of relay. Here the previously neglected processes dominate, namely substrate input at rate V to replenish ATP and secretion to deplete intracellular cAMP. From Figure 6.6 (MDP p108), substrate input must bring about a change in α of about 50 units, *i.e.* , from *(MDP:6.2, p103)*, a change in A of about $50K_S$ units. At constant rate V, the replenishment will take about $50K_S/V$ time units, as compared with the time scale $(k_t^*)^{-1}$ for cAMP secretion. Thus, the parameter $50k_t^*K_S/V$ measures the relative times. A typical value of $k_t^*\ K_S/V$ is 4. Because this dimensionless parameter is not small, it also seems permissible to make a quasi-steady assumption on B and β in the second phase of relay.

[9] Whether it is regarded as decreasing B or increasing C, there is only a single time scale for the secretion process.

Exercise $\aleph 18_6$

Make parameter changes that will still permit a quasi–steady–state assumption on γ but which make that assumption invalid for β. Check your ideas with computer simulations.

Exercise $\aleph 19_6$

By means of "biophysical" reasoning of the type employed above, try to find a combination of parameters that will provide an estimate for the time scales for the decrease and subsequent increase in α as in Figures 6.5 and 6.6 of MDP. Check your reasoning by changing the parameters involved and observing whether computer simulations show the predicted change in the time scales. (Note: The answer to this exercise, given in section 7.4 below, provides the final link in the arguments needed to suggest parameter domains where the quasi–steady–state assumptions on β and/or γ are valid.)

Exercise $\aleph 20_6$

Employ the fact that β changes rapidly compared to α to "explain" the phase plane behavior of Figures 6.12 (MDP, p116) and 6.13 (MDP, p118). In particular, explain why the trajectories sometimes seem either to be roughly horizontal or to follow nullclines. See section 7.4 for the answer.

Comments

The next four exercises have two motivations. The first is to introduce the reader to another well-studied biochemical system that exhibits interesting dynamical behavior. The second is to illustrate the fact that many qualitative conclusions that we have gleaned with the aid of our kinetic models are "robust" (see p. 120 of the text) in that they persist in spite of changes in details of the underlying assumptions.

Goldbeter (1980) summarizes the salient features of the glycolytic oscillations that have been intensively investigated in intact yeast cells, as well as in cell extracts of bacteria, muscle and yeast. In the latter, most commonly studied case, oscillations with a period of about five minutes are observed if constant injection rates of substrate lie between two critical values. The simplest reasonable model involves the catalysis by the enzyme phosphofructokinase (PFK) of the conversion of the substrate adenosine triphosphate (ATP) to the product adenosine diphosphate (ADP). Here we wish to consider the influence of the presumed number of subunits n in the enzyme, as was done by Venieratos and Goldbeter (1979)*. (See the just-cited paper for references to support various assertions made below.)

The enzyme is assumed to be of the Monod-Wyman-Changeux type, as discussed in Chapter 4 of the text and applied in the cAMP model of Chapter 6. Previously we confined ourselves to the case where one of the two conformations (the T state) was entirely unreactive (MDP, p62). Now we permit both R and T states to bind substrate (with dissociation constants K_R and K_T) and to convert it into product (with rate constants k_R and k_T, $k_R \geq k_T$, and corresponding Michaelis constants $K_m^{(R)}$ and $K_m^{(T)}$.) We continue to assume that product regulation occurs solely from binding of product to a site on the R conformation (with dissociation constant K_p).

As in the cAMP model, it is assumed that the substrate is injected into the system at a constant rate V and that the product is removed at a rate proportional (with constant k_t^*) to its concentration. Let A and B denote concentrations of substrate and product. Define dimensionless concentrations by $\alpha = A/K_m^{(R)}$ and $\beta = B/K_p$. Then the model equations are

$$d\alpha/d\tau = 1 - \sigma\phi(\alpha,\beta), \quad d\beta/d\tau = q\sigma\phi(\alpha,\beta) - k_t\beta,$$

where generalizing *(MDP:6.3d, p103)*

$$\phi(\alpha,\beta) = \frac{\alpha(1+\alpha)^{n-1}(1+\beta)^n + L\theta w\alpha(1+w\alpha)^{n-1}}{L(1+\alpha w)^n + (1+\alpha)^n(1+\beta)^n}. \tag{7.21}$$

By analogy with the corresponding definition in *(MDP:6.2, p103)* the dimensionless time is defined by $\tau = t/(K_m^{(R)}/V)$. Again, L is the allosteric constant, σ is the ratio of maximum enzyme activity to V, and $q = K_m^{(R)}/K_p$, $k_t = k_t^* K_m^{(R)}/V$, $\theta = k_T/k_R$, $w = K_m^{(R)}/K_m^{(T)}$. Thus the equations are essentially the same as equations *(MDP:6.3a)* and *(MDP:6.3b)* for substrate and product concentrations in the cAMP model of the text, except that the enzyme activity function ϕ has been generalized to the case of n subunits, with binding and product formation occurring in both the T and R states.

Exercise $\aleph 21_6$

Derive formula (7.21) for $\phi(\alpha, \beta)$.

Exercise $\aleph 22_6$

(a) Consider case "glycosc" (*select* = 9, page 48). Examine the oscillations for $n = 2$ by calling for *glycosc.pic*. In particular, note the period.

(b) Now take $n = 4$. The equations are not nearly so "stiff" in this case, and one can be content with smaller accuracy: thus take *etol* = 0.001. Show that the sole steady state is an unstable spiral (and therefore that there is a good chance that the model still gives oscillations for the same parameter values even though the cyclase is taken to be a tetramer instead of a dimer.)

(c) Use the steady state as initial conditions and integrate for 5 time units. The graph seems empty. Can you explain why? (See Section 7.4 for answer.)

(d) Integrate for half the period observed in part (a), starting at (10,20). You should obtain what seems like a little wiggle in the bottom-left corner of the screen. What's wrong? (See Section 7.4 for answer.)

(e) Reset the scaling so that the screen measures 30 by 30 units. Replot the results of (d).

(f) Make a clean picture of the limit cycle, putting an arrow every time unit. How does the period compare with that for $n = 2$.

(g) A good test for the accuracy of integration is to "run" around the same limit cycle several times. A very accurate integration will repeat almost exactly, but an inaccurate one will show deviations in successive circuits. Try this test on the present limit cycle. Try to improve integration performance by changing integration parameters.

Exercise $\aleph 23_6$

(a) Use case "glycosc" (*select* = 9, page 48) with the basic parameters as set in the program and observe the limit cycle. Also make a "window" plot of the same solution. The calculations are delicate. See *glycosc.key* for one way to obtain the desired results.

(b) Find the nullclines in the phase portrait, thereby observing the same type of S-shaped vertical nullcline that was obtained for the cAMP oscillations.

(c) As was mentioned above, a major experimental fact concerning glycolytic oscillations is that they occur for a "window" of substrate injection rates V. Observe that multiplying V by a factor is equivalent to dividing both dimensionless parameters σ and k_t by the same factor. With this in mind, begin an investigation of the effect of V by respectively multiplying V by 1/2 and 30. Give evidence that in both cases there is a single stable steady state, but that the former case is "excitable", in that a superthreshold stimulus is considerably amplified. Plot the nullclines in both instances. (Note that the S-shaped vertical nullcline is unaltered by a change in V.)

Exercise $\aleph 24_6$

This and the succeeding exercise contain a deeper study of the window in injection rate.

(a) Show that *(MDP:6.27, p114)* is also a necessary and sufficient instability condition for the present model. (More detailed stability results are presented by Venieratos and Goldbeter (1979)*, who of course use these results to guide their choice of parameters for their numerical simulations.)

(b) Use the result of (a) together with information from the previous exercise to explain the observation that glycolytic oscillations exist only for a certain interval of substrate injection rates.

Exercise $\aleph 25_6$

In yeast extracts, sustained oscillations occur when substrate injection varies between 20 and 160 mM/h, *i.e.* injection rate can vary by a factor of about eight. Most of the parameters used in the present simulations were taken from data for *E. coli*, since a complete set of values for yeast were not available when Venieratos and Goldbeter (1979)* performed their analysis. Attempt to verify the following assertion by these authors. For $n = 2, 3, 4, 6$, oscillations occur for values of V that respectively fall between the following multiples of the value of V that is implicit in the standard parameter values of the case "glycosc": 0.22 and 2.5, 0.097 and 1.1, 0.068 and 0.53, and 0.050 and 0.23. Note that the best fit with experiment is when $n = 4$, and indeed the experimental evidence indicates that PFK is a functional trimer or tetramer.

Exercise $\aleph 26_6$

This exercise guides an examination of the waveform of oscillation as a function of protomer number n.

(a) Provide evidence to validate the statement of Venieratos and Goldbeter (1979)* that the value of V that is implicit in case "glycosc" maximizes the amplitude of the oscillation (at 23.8). This amplitude is defined as the difference between maximum and steady state levels of product concentration. It will be sufficient to examine two different values of V, respectively somewhat greater and somewhat less than the standard value.

(b) Use two windows to plot at least three peaks of oscillations, for both substrate and product.

(c) Repeat (b) for $n = 3, 4, 6$ and values of V that are respectively 0.66, 0.33 and 0.12 times the standard value. Make a table of the maximum amplitude and the corresponding period for the four values of n.

The activity of PFK in yeast extract oscillates between 1 and 80 percent of its maximum. Further investigation, not requested here, indeed reveals semi-quantitative agreement with this observation for n

between 2 and 4.

Exercise $\aleph 27_6$

Check the possibility that q might be much larger than unity (as in the cAMP oscillations) by multiplying the q used in case "glycosc" by 100. To obtain oscillations, also multiply k_t by 100. Try $n = 6$ as well as $n = 2$. Plot as in the previous exercise, part (b). The waveforms that you obtain, particularly the sharp spikes when $n = 6$, are quite different from those observed.

Comments

The present decade has seen a explosion of research into the complex dynamical behavior that can be generated by relatively simple equations. A fundamental example has been examined in our discussion of the population *difference* equation *(MDP:2.7, p17)* with its period-doubling and its chaotic solutions. Similarly, the next several exercises demonstrate the rich variety of dynamical behavior that can be generated by *differential* equations of a type with which we are now well familiar.

It is too early to say how much of this intricate behavior is of major importance in biology, but at the very least it is intellectually fascinating. Moreover, it is a reasonable conjecture that complex dynamics will prove of importance at least in one area — neurobiology. Whether the simple and common biological interactions considered here actually are the source of complex behavior in any important biological system is not decisive for our present purposes; we wish merely to illustrate the type of phenomena that might be anticipated. Nonetheless, we shall be able to cite biological examples of several types of behavior that we shall describe.

The model that we shall investigate was proposed and examined by Decroly and Goldbeter (1982)*. It consists of a sequence of two positive feedback loops, each of the type that we have just examined as a model for glycolytic oscillations. In the first loop, a Monod-Wyman-Changeux enzyme E_1 catalyzes the production from a substrate S of a product P_1. Substrate is injected into the system at a constant rate, or is synthesized at a constant rate. The product P_1 activates E_1. All this is identical to the glycolytic model. But in addition, an enzyme E_2

produces a product P_2 from P_1 and is activated by P_2. P_2 decays or is removed. The dimensionless equations of this model are given in case "chaosc", *select* = 10, on page 49.

NOTE: In the following six exercises, the (physiologically reasonable) parameters of case "chaosc" should be used except when changes are specifically requested. (Initial conditions often have to be varied.) When attempting to verify the simulations of Decroly and Goldbeter, keep in mind the remark in section 7.1 (page 126) that there may well be a difference in parameter ranges between their simulations and ours.

Exercise $\aleph 28_6$

(a) Observe the oscillations in case "chaosc" and show that they also occur if k is diminished.

(b) Note that oscillations also occur for large values of the dimensionless removal constant k by trying $k = 35$ and one or two larger values of k.

(c) Try to provide a degree of explanation (as in Decroly and Goldbeter (1982)*, p. 6918) for the behavior found by considering the system in the limits $k \to 0$ (*i.e.* $\gamma \to \infty$) and $k \to \infty$ ($\gamma \to 0$).

Exercise $\aleph 29_6$

Decroly and Goldbeter (1982)* assert that for $3.52 < k < 4.04$ both a stable limit cycle and a stable steady state exist. Check this statement. In particular, try to demonstrate how sufficiently large perturbations can shift the system from one state to the other when $3.5 < k < 4$ (approximately).

Exercise $\aleph 30_6$

(a) Check the Decroly and Goldbeter (1982)* assertion that "birhythmicity" exists for $3.52 < k < 4.04$. Do this by performing simulations for $k = 4$ and initial conditions $\beta = 250$, $\gamma = 0.25$ and $\alpha \approx 32$. Integrate for at least 600 time units. You should observe that the system tends to one of two stable oscillations, depending on the initial conditions on α. Try other initial conditions to demonstrate this point more fully.

(b) In the simulations of (a) you should observe that the system transiently oscillates in a certain manner before it tends to a different final oscillatory mode. Show that the transient behavior is due to the presence of an unstable oscillatory solution; do this by integrating backwards in time.

(c) "Birhythmicity occurs because the model contains two mechanisms for generating instability". Comment.

Exercise $\aleph 31_6$

(a) Test the assertion that for $4.0 < k < 4.4$ (approximately) there is only a single stable limit cycle (Decroly and Goldbeter, 1982)*.

(b) Repeat the simulation of Decroly and Goldbeter (1982)* for about 1800 time units and $k = 4.422$, with initial conditions $\beta = 188.8$, $\gamma = 0.3367$, $\alpha \approx 29.2$. By slightly varying the initial conditions for α, you should observe the coexistence of two possible regimes, one oscillatory and one chaotic.

(c) Decroly and Goldbeter (1982)* found that transition between birhythmicity and a single stable limit cycle occurred at $k = 4.044$. Estimate this critical value of k using the numerical scheme built into BIOGRAPH .

Exercise $\aleph 32_6$

Examine more closely the transition to the chaotic state that was exhibited in Exercise $\aleph 30b_6$. According to Decroly and Goldbeter (1982)* two limit cycles exist for $4.387 < k < 4.422$. The period of one of them keeps doubling, with the initial doubling at $k = 4.404$ and chaos being attained for $4.422 < k < 4.520$. Try to obtain this behavior.

Exercise $\aleph 33_6$

For $4.52 < k < 28.44$, Decroly and Goldbeter (1982)* observe oscillations that are periodic, although complex. Particularly for smaller values of k in this range, they report several cycles in β and γ for a single periodic variation of α. Simulate some examples of this behavior.

Exercise $\aleph 34_6$

In a second paper, Decroly and Goldbeter (1984)* demonstrate another aspect of the model under investigation, a complex dependence of behavior upon initial conditions.

(a) Test the existence of birhythmicity for the standard parameters except that $\sigma_1 = \sigma_2 = 25.64$, $k = 4.10$. Take initial conditions $\beta = 128$, $\gamma = 0.244$, $\alpha = 35.6$ or 35.7 and integrate for 800 time units. Do you observe a transient chaotic nature of the solutions as did Decroly and Goldbeter? What happens if parameters are varied slightly?

(b) Decroly and Goldbeter (1984)* systematically varied the initial value of α, taking finer and finer steps to traverse ever diminishing intervals: (34, 60), (44, 46), (44.3, 44.5), (44.40, 44.42). At every level of detail they found alternating subintervals of points starting from which the solution tended first to one and then to the other limit cycle. Try to demonstrate this fine structure by examining initial values of α between 44.411 and 44.415. Many time-consuming integrations have to be carried out; it is thus advisable to work on the simulations in parallel with other studies, readings, or amusements.

If there are two stable asymptotic states, the initial conditions determine which will be attained. The simplest assumption is that there is a smooth surface that separates initial conditions that yield one or the other of the two possible final states. The present calculations show, however, that this surface may be infinitely convoluted; for its structure appears to be of so-called "fractal" nature, wherein the same level of complexity manifests itself at all levels of detail.

(c) Basing their idea in part on the chaotic transient mentioned in (a), Decroly and Goldbeter suggest that the separating surface may be an unstable "strange attractor". (We are familiar with ordinary attractors that are constituted of a single point (steady state) or a closed curve (limit cycle). Chaotic solutions typically tend to a much more complicated "strange" surface; their wanderings on this surface are responsible for their seemingly random behavior.) Check for the presence of such an unstable attractor by integrating backwards in time.

Comments

Let us briefly discuss the results of the last set of exercises. See Decroly and Goldbeter (1982)* and the review by Goldbeter, Martiel, and Decroly (1984)* for further discussion and references.

One question that at once arises is: What evidence is there that the phenomena exhibited by the model actually occur and/or are of relevance in biology? In partial answer, there are observations of complex periodic oscillations, bistability, and chaos in chemical and biological systems. In particular, the former two phenomena have been observed in yeast glycolysis with constant input, while chaotic dynamics in yeast glycolysis with periodic inputs of substrate have recently been the object of extensive theoretical and experimental study. See Markus and Hess (1984)* and Markus, Kuschmitz and Hess (1984)*. The text sketched some results concerning the periodic forcing of a naturally periodic system in its treatment of stimulated heart cells (pp. 28–31, MDP). Complicated dependence on initial conditions can be found in a study of epidemics by Schwarz (1985)*

Irregular cAMP signaling has twice been observed in slime mold mutants, while chaotic dynamics has been predicted in a new theoretical model of cAMP signaling that incorporates receptor desensitization (Martiel and Goldbeter, 1984; Goldbeter *et al*, 1984)*. In the model studied in Exercises $\aleph 28_6$—$\aleph 34_6$, the complex behavior was generated by the serial linkage of two instability-generating mechanisms, in each of which positive feedback of the product on the catalyzing enzyme competes with negative feedback by substrate depletion. In the new cAMP signaling model, one negative feedback of the above type exists, but in parallel there is negative feedback caused by passage of the receptor into an inactive state.

Exercise $\aleph 35_6$

Here is an example showing how observation of computer solutions carried out by the authors in the course of preparing these exercises) can lead to new insights that can be verified analytically.

(a) The authors tried to see how large σ could be raised, from its value of 30 in Figure 6.15 of MDP p25, before relay ceased. Relay was found at

$$\sigma = 30.5,\ 31,\ 32,\ 35,\ 40,\ 50,\ 100,\ 200,\ 500,\ \textit{and}\ 1000$$

Verify that indeed relay with a sharp threshold exists at one or two of these values of sigma.

(b) Take a step toward analytic explanation of these results by showing that for large σ the steady state is just to the left of the maximum point of the vertical nullcline.

This problem demonstrates that relay can sometimes occur over broad parameter ranges.

Exercises for chapter 7 of MDP

We wish to use the computer to solve the diffusion equation

$$\frac{\partial a}{\partial t}(x,t) = D_a \frac{\partial^2 a}{\partial x^2}(x,t) \quad \textit{for} \quad 0 \le x \le L, \quad 0 \le t\,. \tag{7.22}$$

To do so, we approximate x-derivatives by difference quotients, as follows:

$$\frac{\partial a}{\partial x}(x,t) = \frac{1}{\delta}\left\{a(x+\frac{1}{2}\delta,t) - a(x-\frac{1}{2}\delta,t)\right\} \tag{7.23}$$

$$\begin{aligned}\frac{\partial^2 a}{\partial x^2}(x,t) &= \frac{1}{\delta}\left\{\frac{\partial a}{\partial x}(x+\frac{1}{2}\delta,t) - \frac{\partial a}{\partial x}(x-\frac{1}{2}\delta,t)\right\} \\ &= \frac{1}{\delta^2}\left\{a(x+\delta,t) - 2a(x,t) + a(x-\delta,t)\right\}.\end{aligned} \tag{7.24}$$

Let the interval $(0, L)$ be divided into M subintervals of length δ. If we make the abbreviation

$$a(i\delta,t) \equiv a_i(t) \ , \quad i = 1, 2, ..., M \tag{7.25}$$

then we are led to the approximation of (7.22) by a system of ordinary differential equations

$$\frac{\mathrm{d}a_i}{\mathrm{d}t}(t) = \frac{D_a}{\delta^2}\left\{a_{i-1}(t) - 2a_i(t) + a_{i+1}(t)\right\} . \tag{7.26}$$

Such a system is of the type that BIOGRAPH has been designed to solve. Because it is not within our scope to probe deeply into the mysteries of partial differential equations, we shall only consider the special case wherein the M subintervals are arranged in a circle. Under this assumption, the diffusion effectively takes place in a thin ring-like domain of space; x measures the circumferential distance around this ring.

For diffusion in a ring, subinterval 1 has subinterval M as its left neighbor. Thus the system is described by (7.23) for $i = 2, 3, ..., M-1$. In addition

$$\frac{\mathrm{d}a_1}{\mathrm{d}t}(t) = \frac{D_a}{\delta^2}\left\{a_M(t) - 2a_1(t) + a_2(t)\right\} \tag{7.27}$$

and

$$\frac{\mathrm{d}a_M}{\mathrm{d}t}(t) = \frac{D_a}{\delta^2}\left\{a_{M-1}(t) - 2a_M(t) + a_1(t)\right\} \tag{7.28}$$

The case *select* = 19 has been designed to solve certain equations for two chemicals, a and h, that react as well as diffuse. If we set the reaction terms equal to zero[10], then we can simultaneously solve (7.22) and the entirely analogous equation

$$\frac{\partial h}{\partial t}(x,t) = D_h \frac{\partial^2 h}{\partial x^2}(x,t) \ \ for \ \ 0 \le x \le L, \ \ 0 \le t \ . \tag{7.29}$$

[10] by setting the parameters ρ_h, ρ_a, μ_h, μ_a all equal to zero

Equation (7.29) is approximated by the system

$$\frac{dh_i}{dt}(t) = \frac{D_h}{\delta^2}\left\{h_{i-1}(t) - 2h_i(t) + h_{i+1}(t)\right\} . \tag{7.30}$$

with special equations for $i = 1, M$ as in (7.27) and (7.28). We can thus employ *select* = 19 to check two different parameter regimes at once. The cases ring4 and ring8 treat approximations to the pair of diffusion equations (7.30) and (7.26) for $M = 4$ and $M = 8$ respectively. Note that you must set the parameters multiplying the reaction terms to zero.

The object of the next exercise is to check the "rule of thumb for diffusion" that was discussed on page 140 of MDP. We shall consider the rule in the form *(MDP:7.21, p141)* : material diffuses a distance L in a time of magnitude L^2/D, where D is the diffusivity.

Exercise $\aleph 1_7$

(a) Suppose that a quantity of small molecules is placed at the "north pole" (*i.e.* in cell number 1) of a circle of tissue whose circumference is $2L$. For a first approximation, assume that the tissue may be assumed to be essentially aqueous in nature. In this case we may take D to be of a magnitude suitable for the diffusion of a small molecule in water, i.e. $D = 10^{-5}$ cm^2/sec. Let $L = 100\mu m$. Show how BIOGRAPH can be efficiently used to check whether the rule of thumb gives a reasonably accurate estimate of the time it takes a substantial quantity of the molecules to arrive at the "south pole". Start by examining the effect of doubling the ring circumference. Note that by using two different values for D_h and D_a in each "run", you can explore the dependence of the diffusion time upon the diffusivity.

(b) Should ring4 or ring8 give results that are closer to the "rule". Why?

If you need a little help, see page 220 for a partial answer to this exercise.

Exercise $\aleph 2_7$

The diffusing molecules may be large and/or the medium may be much more viscous than water. In these cases a smaller value of D would be appropriate. On the other hand, a cellular medium may be subject to gross but rather random streaming motions. The effect of these motions might be modeled by making the diffusivity larger. Thus, investigate the consequences of multiplying the diffusivity D used in the previous exercise by a factor of 10, and of dividing D by the same factor.

Exercise $\aleph 3_7$

(a) Instead of a ring of M subintervals, let us consider a ring of M identical cells. Imagine that cellular streaming or some other mechanism keeps the concentration of chemicals uniform within a given cell. Assume that a certain chemical leaks between cells at a rate that is proportional to the concentration difference between adjacent cells. (Compare the first complete paragraph on p103 of the text and also the solution of Exercise $\aleph 13_6$.) Show that a system of differential equations such as (7.26) and (7.29) is a suitable model in this case, if D_a/δ^2 is regarded as a permeability (which we shall denote by k_a).

(b) How long does it take a chemical to permeate M cells? This question is related to the "rule of thumb" for diffusing a distance L. In the present case, however, no physical distance enters, just the number of cells, since each cell is assumed to be perfectly stirred. What would you propose as a "rule of thumb" here? Employ three approaches in suggesting an answer.

- Use dimensional reasoning.
- Formulate and solve a problem for just two cells.
- Modify the "rule" for diffusing a distance L.

Check your conjecture by comparing results for 4 and 8 cells, for a fixed permeability k_a. Then vary the permeability.

See page 221 for a partial answer.

Exercise $\aleph 4_7$

Suppose that a chemical is initially present not only in the subinterval at the "north pole" but also in equal amounts, in the two adjacent subintervals (or cells). What now should be the time for an appreciable amount of chemical to accumulate at the "south pole"? Use the computer to check your conjectures. Gierer–Meinhardt activator–inhibitor simulations

Exercises for chapter 8 of MDP

Exercise $\aleph 1_8$

(a) Make a plot using the standard parameter values of case *ring4.par*. Observe that in each segment the chemical concentrations become effectively constant (*i.e.* temporal variation ceases eventually). Employ F6≻[f] to demonstrate the existence of spatial pattern.

(b) Replot the results of (a) in two windows so that all inhibitor concentrations are displayed in window 1 and all activator concentrations in window 2.

(c) Make a plot using the standard values of case *ring8.par*. Should you have anticipated the existence and nature of the transient oscillations? Extend the computation until you are quite sure that a steady state has been attained.

(d) In either or both of the cases, see what different final patterns, if any, you can generate by varying the initial conditions.

Exercise $\aleph 2_8$

(a) Show that the graph yielded by the $select = 19$ case "giermein" on page 56 is consistent with the prediction of Equation *(MDP:8.19, p153)*, since (as should be verified) $Q = 0.35294$, $R = 4.2857$.

(b) Show how "giermein" can be used to deal with the equations in their simpler dimensionless form *(MDP:8.10)* and *(MDP:8.11)*. Use the computer to locate the steady state point and to determine its stability. Compare the results with analytic predictions.

(c) Continuing part (b), sketch out the main features of the phase plane.

Exercise $\aleph 3_8$

This exercise concerns a comparison between linear stability theory and simulations of the nonlinear equations *(MDP:8.10)* and *(MDP:-8.11)* in situations where linear theory predicts an oscillatory instability.

(a) Consider $R = 4.2857$ (as in the previous exercise) and examine Q as it decreases in steps of 0.1 from 0.7 to 0.1. Plot the results as functions of time. For the periodic solutions, measure the distance between peaks with a ruler and hence – inelegantly but rapidly – obtain the periods.

(b) Compare the periods found in (a) with those predicted by linear stability theory. In what situations are the linear predictions more accurate? Why should such a trend be anticipated? (See page 221 for the answer.)

Exercise $\aleph 4_8$

The purpose of this exercise is to give a simple demonstration of the fact that the Gierer-Meinhardt equations *(MDP:8.1, p150)* are capable of yielding inhomogeneous spatial patterns providing inhibitor diffusion is sufficiently more rapid than activator diffusion.

(a) Read in ***ring4.par***, integrate, and make the plot. Verify that the graphs show a nonuniformity in inhibitor concentration. (Note that four curves appear at the beginning of the plot but only three later, since in two (non-adjacent) cells the inhibitor concentrations tend toward the same value.) Employ F3≻[m] to make the window map 0100 and thereby to exhibit the activator concentrations.

(b) Change D_a to a value only slightly less than D_h, integrate again, and observe that now the solution tends to spatial uniformity.

(c) Calculate P, Q, R and U. Verify *(MDP:8.27, p155)*. Also rationalize the observed pattern with the aid of *(MDP:8.42, p159)*.

(d) Read in *ring8.par*, integrate, and make the plot. Change the window map to exhibit the activator concentrations. Make a rough sketch of the spatial dependence of the activator concentration, employing F6≻[f] for numerical values. Integrate further, until F6≻[e] indicates that the solutions have nearly reached steady states.

(e) Repeat (b) for the case *ring8.par*.

(f) Repeat (c) for the case *ring8.par*.

Exercise $\aleph 5_8$

It will be useful to prepare ourselves for situations wherein there is an imposed spatial variation in the source strengths ρ_a and ρ_h of activator and inhibitor. The activator and inhibitor concentration scales of *(MDP:8.7, p151)* are no longer appropriate, because ρ_a and ρ_h will vary from segment to segment.

(a) Use new dimensionless variables

$$A = a/(c_a/c), \ H = h/(c_a/c) \tag{7.31}$$

Show that the system *(MDP:8.10)*, *(MDP:8.11)* is replaced by

$$\frac{\partial A}{\partial \tau} = S(1 + \frac{A^2}{H}) - A, \qquad \frac{\partial H}{\partial \tau} = WA^2 - QH, \tag{7.32}$$

where

$$S \equiv \frac{\rho_a c}{\mu_a}, \qquad W = \frac{\rho_h c_h c_a}{\mu_a c}, \qquad Q = \frac{\mu_h}{\mu_a} \tag{7.33}$$

Note that case "giermein" can be employed if one sets

$$\rho_a = S, \qquad \rho_h = W, \qquad \mu_h = Q,$$

and all other parameters are set equal to unity.

(b) There are now three dimensionless parameters, in contrast to the two found in *(MDP:8.11)*. Explain. (See Section 7.4 for answer.)

(c) Show that the steady states of *(MDP:8.14, p152)* are replaced by

$$A = (1 + \frac{Q}{W})S, \qquad H = \frac{W}{Q}(1 + \frac{Q}{W})^2 S^2, \tag{7.34}$$

and that instead of the stability results *(MDP:8.18)* we now have

$$\gamma = Q, \qquad \beta = \frac{W - Q}{W + Q} + Q. \tag{7.35}$$

(d) Write $\beta^2 - 4\gamma$ in the form of a quadratic in W whose coefficients are functions of Q. Describe how you would find all domains in the (W, Q) parameter plane for which linear stability theory predicts that the steady state is an unstable focus – but do not carry out the calculations.

(e) We wish to find several examples where linear theory predicts unstable foci. Instead of carrying out the full program of part (d), we will do only a little further analytical work (by seeking easy special cases) and then let the computer help us.

The instability condition $\beta < 0$ yields

$$W < \frac{Q(1 - Q)}{1 + Q}. \tag{7.36}$$

One simple possibility is to take $Q << 1, \quad W < Q$. Show that in this situation one obtains an unstable node, not an unstable focus.

Another possibility is to take $W \ll Q$, in which case $\beta \approx Q - 1$. Show that all required conditions should be fulfilled in this case if $0.17 < Q < 1$.

Now use the computer for $S = 2.353$ and $Q = 0.8$. Find several oscillatory solutions for sufficiently small values of W. Note that we have shown that $W \ll Q$ necessarily leads to solutions of the type desired, but larger values of W are not ruled out — and should be attempted.

(f) If you have not done Exercise $\aleph 3_8$, use the procedure described therein to compare the predictions of linear theory with the results obtained for the oscillation periods when the nonlinear equations are integrated for the (Q, W) pairs of part (e).

Comments

The remaining exercises for this chapter involve the numerical solution of the partial differential equations *(MDP:8.1, p150)* that exemplify the Gierer–Meinhardt approach to pattern formation in developmental biology. Some discussion is required before we proceed.

BIOGRAPH numerically solves the system of ordinary differential equations (3.72), (3.73). These are essentially the same as *(MDP:8.50, 8.51, p161)* except that instead of the difference quotients

$$\frac{a_i(t+\Delta t) - a_i(t)}{\Delta t} \qquad \textit{and} \qquad \frac{h_i(t+\Delta t) - h_i(t)}{\Delta t}$$

equations (3.72) and (3.73) employ the derivatives da_i/dt and dh_i/dt. Where *MDP* employs Δx we now employ the symbol δ. (The computer's name for this parameter is "deltax".)

MDP presented the simplest reasonable numerical scheme for solving the Gierer–Meinhardt equations *(MDP:8.1a, 8.1b, p150)*. We now approximate such partial differential equations by a system of ordinary differential equations, and then use BIOGRAPH's rather sophisticated scheme for solving such systems. This approach to the numerical analysis of partial differential equations is called the *method of lines.*

Partial differential equations are typically solved in some region of space. The equations must be supplemented by extra conditions that describe what is happening at the boundaries of these regions. The Gierer–Meinhardt equations, for example, might be regarded as holding for some line of tissue that extends from $x = 0$ to $x = L$. If no chemical flows through the boundary of this tissue, then it follows from *(MDP:-7.7, p134)* that $\partial a/\partial x$ and $\partial h/\partial x$ must vanish at $x = 0$ and $x = L$. We have used, however, boundary conditions appropriate for a spatially periodic domain. We can interpret such a domain as a ring of cells without boundary. The reader should consult the Comments on page 174 for further details about how we treat the ring of subintervals that comprise our spatial domain.

Usually it will be convenient, as in MDP, to treat the equations in their dimensionless form *(MDP:8.8)*. The latter equations can be regarded as a special case of *(MDP:8.1)* with all parameters unity except that

$$c = R, \qquad c_h = Q, \qquad \mu_h = Q, \qquad D_h = P.$$

We consider a ring of n subintervals, each of width δ. Thus all our solutions have period $n\delta$. As usual, we need theoretical guidance as to what types of solutions to expect as parameters vary. For this we shall rely on *(MDP:8.24, p155)*, a necessary and sufficient condition for the stability of perturbations proportional to $\cos(s/l)$. If the inequality is reversed, we have an instability condition.

Bear in mind that *(MDP:8.24)* should not be expected to provide quantitatively accurate guidance. It deals with perturbations that are small in amplitude and sinusoidal in form. Moreover, it refers to partial differential equations, while the simulations deal with a discretized translation into a system of ordinary differential equations. The two sets of equations are expected to have solutions that become identical only as the number of cells $\longrightarrow \infty$ so that $\Delta x \to 0$.

BIOGRAPH does not provide the best possible graphical facilities for observing spatial patterning. One good method, however, is to plot all or a sample of activator concentrations in one window, with corresponding inhibitor concentrations in another window. Monitoring the plots as they are graphed gives an excellent feeling for the dynamics of the pattern formation. (In particular one can observe the not-infrequent cases where two or more cells come to behave identically.) A useful tactic in exploratory calculations is to dispense with time-consuming plots altogether, by means of F3≻[m] or perhaps F5≻[i]. One can then employ F6≻[f] to examine the nature of the pattern. F6≻[e] is useful for determining whether a steady state has been attained. Moreover, this command displays even and odd subscripted variables in two separate columns, making it easy to discern trends in activator and inhibitor concentrations.

Exercise $\aleph 6_8$

This exercise concerns the one-dimensional version of the equations and parameters used to obtain Figure 8.6 on p163 of MDP. The major goal is to discuss the nonuniqueness that is depicted in the figure, where different "past histories" lead to different final states for identical final equations and boundary conditions.

(a) Using the parameters in the caption of Figure 8.6, show that

$$P = 50, \qquad Q = 1.33333, \qquad R = 6.66667$$

and that the spatially uniform steady states (which give an idea of the scaling to try for dependent variables) are

$$\overline{A} = 7.66667, \qquad \overline{H} = 58.77778.$$

(b) Show that in this case *(MDP:8.24, p155)* predicts instability of solutions proportional to $\cos(s/l)$ if $1.2 \leq l \leq 5.0$. Bear in mind that a sinusoidal solution of period $n\delta$ corresponds to a situation wherein $n\delta = 2\pi l$. Using this, show that *(MDP:8.24)* predicts that sinusoidal perturbations of period $2\delta, (4/3)\delta$, and 4δ will *all* grow for $3.8 \leq \delta \leq 5.9$.

(c) Let us consider $n = 4$. If linear stability theory is a guide, then there is a range of "cell widths" δ which can be expected to yield solutions with 1, 2, or 3 peaks. To test this, set $\delta = 5.75$ and try the initial conditions

$$y_1 = y_3 = y_5 = y_7 = 3.5, \qquad y_2 = 1.2, \qquad y_4 = y_6 = y_8 = 1.3,$$

and

$$y_1 = 7,\ y_3 = 8,\ y_5 = 7,\ y_7 = 8,\ y_2 = 2,\ y_4 = 3,\ y_6 = 2,\ y_8 = 3.$$

Attempt to obtain the missing anticipated final solution by imposing suitable initial conditions. Record at least one of the final results for use in Exercise $\aleph 5_8$, part (d).

Exercise $\aleph 7_8$

(a) Continue the previous exercise, but now take $n = 8$. Try the following 3 sets of initial conditions for $y_1, y_2, \ldots, y_{16}$.

$$80,\ 10,\ 40,\ 5,\ 80,\ 10,\ 40,\ 5,\ 80,\ 10,\ 40,\ 5,\ 80,\ 10,\ 40,\ 5 \qquad (7.37)$$

$$80,\ 10,\ 80,\ 10,\ 40,\ \ 5,\ 40,\ \ 5,\ 80,\ 10,\ 80,\ 10,\ 40,\ \ 5,\ 40,\ \ 5 \tag{7.38}$$

$$80,\ 10,\ 80,\ 10,\ 80,\ 10,\ 80,\ 10,\ 40,\ \ 5,\ 40,\ \ 5,\ 40,\ \ 5,\ 40,\ \ 5 \tag{7.39}$$

Record the steady solutions obtained.

(b) In each case of (a) perturb the final solution by changing component 1 by about 25%. What happens?

(c) Draw crude graphs of the different steady solutions that you found. Can you generate any more?

Exercise $\aleph 8_8$

For the steady solution resulting from initial conditions (7.39) of the previous exercise use $n = 4$ to simulate the results of cutting the 8-interval ring in half and rejoining the ends. If you have time, check all possible different situations. Try to predict the form of the steady states that you will obtain.

Exercise $\aleph 9_8$

Simulate "growth" by taking one of the final solutions found in Exercise $\aleph 6_8$, part (c), and adjoining two cells on either side with activator and inhibitor concentrations set at their steady–state values. Try to find interesting behavior for other values of δ . For which, if any, do you recover the result of *MDP* Figure 8.6 that activator peaks are more widely separated in growing tissue?

Exercise $\aleph 10_8$

Exercise $\aleph 6_8$ explored the behavior of solutions when $\delta = 5.75$. Try other values of δ, first predicting behavior analytically and then testing it. Start by predicting values of δ for which only the uniform steady state is expected to be a final state, regardless of initial conditions.

Comments

A major phenomenon demonstrated by the previous exercises is the formation of activator peaks in a circle of tissue, by means of an instability of an initially near-uniform chemical state. The number of peaks varied according to the conditions. A study of the cellular slime mold *Polysphondylium pallidum* by Byrne and Cox (1986)* provides some of the best experimental evidence to date of the biological relevance of this finding. The life cycle of *P. pallidum* is very similar to that of *D. discoideum* (see MDP, p97). However, instead of a single spherical mass of spores atop a stalk, the fruiting body of *P. pallidum* is composed of sets of branches, one to eight in each set. The several sets are regularly spaced along the central vertical stalk. The branches radiate horizontally, perpendicular to the central stalk. Within each set, the angles between branches are approximately equal, and each branch is ultimately tipped with a mass of spores.

The formation of the sets of branches commences with the segregation of regularly spaced spherical masses (or "whorls") along the central stalk. One to eight tips form along the equator of the each whorl and appear to organize further development. Byrne and Cox (1986)* provide detailed information concerning tip formation by means of three tip-specific monoclonal antibodies. They report that "prepatterning of secondary tips begins with a homogeneous activation of the cells on the surface of the whorl (which)...becomes localized as an equatorial band... The equatorial band of tip antigens breaks up into clusters that form a prepattern for secondary tip formation."

These events are well modeled by the formation of a number of activator peaks in a ring of tissue. Indeed, Byrne and Cox specifically state that the pattern they observe may well arise from a symmetry-breaking mechanism of the reaction-diffusion type (that is the subject of MDP's Chapter 9) or of a type based on mechanical forces (as is discussed in Chapter 9 of MDP).

Exercise $\aleph 11_8$

This exercise provides an introduction to the possibility that intracellular communication can synchronize individual cellular oscillators with different frequencies.

(a) Employ case *ring4.par*, with the diffusivities set equal to zero, simultaneously to exhibit oscillations in activator and inhibitor concentrations with four different frequencies. To this end, in the equations of (7.32) set the parameter values at $S = 2.353$ (as in Figure 8.7 of MDP), $Q = 0.8$, and $W = 0.01,\ 0.02,\ 0.03,\ 0.04$, respectively, in the four "cells". Start the simulation with initial conditions $y_1 = 7,\ y_2 = 8,\ y_3 = 9, \ldots, y_8 = 15$. Measure the four different periods, perhaps simply by marks on a piece of paper.

(b) Now set $D_a = D_h = 1$, to simulate diffusive coupling of a ring of four cells. What do you anticipate will happen to the four periods of oscillation? Compare your expectation with the simulation results.

(c) Try $D_a = D_h = 10$ and also try $D_a = 1$, $D_h = 100$, with the same initial conditions as in part (a). Conjecture the results before running the integrations.

(c) Try to obtain qualitatively different results by varying the initial conditions.

Comments

Biological variability is such that no two oscillators in a cellular population will have exactly the same frequency. The coupling of cellular oscillators is seen in a variety of circumstances ranging from an assemblage of oscillating *Dictyostelium* cells to a beating heart. For further reading see Winfree (1980) or Ermentrout (1985)*.

Exercise $\aleph 12_8$

We wish to reproduce the results of Figure 8.10 (*MDP, p172*) showing that the "tissue" will automatically generate a gradient, and that this gradient will remain even when a portion of the tissue is removed. With BIOGRAPH we cannot directly impose the required zero-flux boundary conditions *(MDP:8.53, p165)*. Nonetheless we can proceed indirectly by considering only the portion of the pattern that we generate that falls between two extrema. (The derivatives will vanish, as required, at the ends of this portion.) We shall proceed in this spirit.

(a) Show that the dimensionless parameters corresponding to Figure 8.10 are

$$R = 25, \quad P = 56.25, \quad Q = 2.5.$$

Use case "giermein" to verify that the uniform state is stable to spatially uniform perturbations, as shown in Figure 8.10.

(b) Use case "ring8" and take deltax=2. Show that various initial conditions give a distribution of activator and inhibitor that is roughly sinusoidal in shape. The portion between the maximum and the minimum can thus be regarded as a representation of the gradient pattern that appears in Figure 8.10.

(c) Now take deltax=1.5, which is equivalent to shortening the length of the tissue by 25%. Take the final values of part (b) as initial conditions (employing F4≻[f]) and integrate until a new steady state is attained. Show that the gradient pattern is maintained.

Exercise $\aleph 13_8$

This exercise is concerned with the parameter dependence of the onset of spatial inhomogeneity. We shall employ the dimensionless variables of (7.31) instead of those defined in *(MDP:8.4, p151)* and *(MDP:8.7)*. We will thus replace *(MDP:8.8)* by corresponding equations wherein the reaction terms are those of (7.32), that is

$$\frac{\partial A}{\partial \tau} = S\left(1 + \frac{A^2}{H}\right) - A + \frac{\partial^2 A}{\partial s^2}, \qquad \frac{\partial H}{\partial \tau} = WA^2 - QH + P\frac{\partial^2 H}{\partial s^2}. \tag{7.40}$$

(a) Consider equation (7.40) with

$$P = 50, \quad S = 1, \quad Q = 1.33333. \tag{7.41}$$

Check these analytic results.

- From (*MDP:8.24*), $U = 0.026667$.
- From (*MDP:8.33*), $R_c = 2.0925$, hence $W_c = 0.6372$.
- From (*MDP:8.29*), instability is predicted to set in for $\ell \equiv \ell_c = 2.475$.
- If 4 cells of width Δx are to contain a half-wave as in MDP Figure 8.8(a), then $\delta = \mathit{deltax} = 1.9444$.

(b) The analytic calculations give an idea of interesting parameters. To estimate the actual boundaries of stability for the four equations of case *ring4.par*, start with the initial conditions

$$y_1 = 5, \quad y_2 = 3.5, \quad y_3 = 4.8, \quad y_4 = 3.3, \quad y_5 = 4.6, \quad y_6 = 3.1,$$

$$y_7 = 4.4 \quad y_8 = 2.9.$$

Make the parameter settings corresponding to (7.41):

$$D_h = 50, \quad rhoa = 1, \quad muh = 1.33333, \qquad (7.42)$$

with

$$rhoh = 0.6, \qquad deltax = 1.944, \qquad (7.43)$$

and all other parameters equal to unity. (As usual in this case, the parameters $rh1, rh2, \ldots, ra1, ra2, \ldots$ should be ignored unless they are specifically mentioned.) Use F5≻[m] to integrate for 30 time units, and show that no inhomogeneity is evident.

(c) Keeping the same initial conditions, try $W \equiv rhoh = 0.4$ and 0.2, and thereby show that only in the latter case is inhomogeneity clearly present. When $W = 0.4$ there is very little inhomogeneity after 30 time units, but there is some — particularly in the activator concentration. To dispel any suspicion that inhomogeneous solutions appear when $W = 0.4$ integrate for a sufficiently long time so that all concentrations are uniform to three significant figures.

(d) By further integrations, compare the nature of the inhomogeneity for $W = 0.2$, 0.15, 0.1 and 0.05. You should find progressively more prominent activator peaks.

(e) In (c) it was determined numerically that the critical value of W for the onset of inhomogeneity lies between 0.4 and 0.2. Establish this value more accurately. You should find that lengthy integrations are required to yield the final pattern when W is only slightly below its critical value.

Exercise $\aleph 14_8$

Set $W = 0.2$. Investigate the effect of initial conditions on the orientation of the maximum in activator and inhibitor concentration.

Exercise $\aleph 15_8$

The stability theory in the text can easily be adapted to the form (7.40) of the Gierer-Meinhardt equations merely by replacing R by Q/W. The dimensionless activator strength S is predicated to be without influence. Nonetheless, perhaps S may affect large disturbances, not covered by the small perturbation theory. Check this point by redoing part (c) of the previous exercise when $S \equiv rhoa \equiv 5, 25, 250$, and 2500.

Exercise $\aleph 16_8$

In this exercise we study the effects of spatial nonuniformity in source distribution. The appropriateness dimensionless equations are those of (7.40) wherein $W \equiv rhoh$ is the dimensionless strength of inhibitor sources and $S \equiv rhoa$ is the corresponding quantity for activator sources. Exercises $\aleph 13_8$ $\aleph 15_8$ have indicated that when S and W are uniform, W but not S effects the onset of inhomogeneity in activator and inhibitor concentrations. We thus expect that the major effects of nonuniformity will arise from a spatial distribution of inhibitor sources. Exercise $\aleph 13_8$ demonstrated that close to critical conditions for the onset of spatial inhomogeneity not only are inhomogeneities weak but they develop very slowly. We thus will center our investigations upon the rather strongly unstable situation where the parameters are as in (7.41), (7.42), and (7.43) but with $W = rhoh = 0.1$.

(a) Carry out the calculations as in part (b) of Exercise $\aleph 13_8$, and write down the activator and inhibitor concentrations when they have reached a fairly steady state.

(b) By varying the initial conditions, try to obtain patterns wherein activator and inhibitor peaks are shifted compared to the results of (a).

(c) Establish a nonuniformity in inhibitor source by setting $rhoh$ equal to a negative number and

$$rh1 = 0.09, \qquad rh2 = 0.1, \qquad rh3 = 0.11, \qquad rh4 = 0.1.$$

Note that the average strength of the inhibitor sources is the same as in part (a). Carry out the integration with the same initial conditions as in (a). You should observe results similar to those of part (a).

(d) To show that the pattern will appear even if virtually no activator and inhibitor are initially present set the initial conditions to values $m \cdot 10^{-3}$, where m is an integer that you choose "randomly" for the different variables. Integrate until a steady state is attained.

Exercise $\aleph 17_8$

To study the possible role of slight asymmetries in orienting a pattern, compare the relative effects of initial conditions and source inhomogeneity in determining the nature of the final steady state. For example, given that there is a certain peak in inhibitor source strength in segment 1, how strong an initial inhibitor peak in segment 3 (if any) will yield a peak in segment 3 at steady state?

Comments

Analysis of the Turing-Gierer-Meinhardt models seeks conditions where very small inhomogeneities are magnified, so that a final inhomogeneous steady state can be anticipated. The (false) impression sometimes arises that these models (and similar ones) are based on the assumption that spatial pattern emerges from a situation in which no pattern at all was initially present, only slight random initial inhomogeneities. It is true that these models *can* explain such an origin of pattern, and there are indeed biological situations where this type of explanation is relevant. Nonetheless, embryos generally have built-in orientations. The sea urchin embryo exhibits both possibilities (Meinhardt 1982, p24). The animal-vegetal axis is fixed during egg formation, but minor external perturbations can shift the orientation of the dorsoventral axis.

Exercise $\aleph 14_8$ dealt with an example of pattern reorientation by initial conditions in a situation devoid of built-in asymmetry, while Exercise $\aleph 17_8$ concerned the ability of a preexisting small inhomogeneity to establish the orientation of a pattern in the face of inevitable random fluctuations.

It is worth emphasizing that analytic testing of the stability of a homogeneous state is helpful even when inhomogeneity is built in, for the general results of the theory should be valid if the underlying nonuniformities are not too strong. An example of this fact is given on p163 of MDP in connection with Figure 8.7, a situation where quite regularly spaced activator and inhibitor peaks are obtained in the presence of spatial gradients in source distributions. Conditions guaranteeing the appearance of these peaks can be quite well approximated by replacing the actual source strengths by their (constant) averages.

One more point is worth stressing. Suppose that there is an underlying weak nonuniform source distribution. Exercise $\aleph 16_8$ shows that such a nonuniformity can be amplified greatly when conditions are such that a uniform state is strongly unstable. Our calculations thus demonstrate how a weak "prepattern" can be inherited and then later can become amplified in the course of development. See Meinhardt (1982) for many additional highly informative computer simulations based on the approach to developmental pattern via reaction-diffusion equations.

Exercises for chapter 9 of MDP

Exercise $\aleph 1_9$

Employ the *select* = 11 case on page 51.

(a) Use the case "catrig" to reproduce Figure 9.12(c), p.212 of MDP, setting $\alpha' = 0.4$. (N.B. The solution to this exercise is worked out in the file *bistab.key*.)

(b) Change the parameter α' to 0.33, and reproduce Figure 9.12(a).

(c) With $\alpha' = 0.384$, reproduce Figure 9.12(b).

(d) With $\alpha' = 0.51$, reproduce Figure 9.12(d).

Exercise $\aleph 2_9$

Try to choose parameters that will reproduce, at least roughly, the graphs of Figure 9.14 on page 217 of MDP. A solution is provided by the *focal.par* file.

Exercise $\aleph 3_9$

Draw trajectories in the phase plane for a situation where there are only two steady states, both stable nodes. Thereby show that it seems unavoidable to assume in addition that there is a third steady state, a saddle point. This brings one to Figure 9.4 on p203 of MDP. More advanced texts on differential equations provide theorems that often allow one to pass from a conclusion that "it seems unavoidable" to theorems concerning qualitative behavior in the phase plane.

Exercises for appendix 1 of MDP

Exercise $\aleph 1_{A1}$

(a) Assign parameter values in the case "acubic" ($select = 5$, page 37) to find and graph the solution to

$$dx/dt = x, \qquad x(0) = 1; \qquad 0 \leq t \leq 3.$$

Obtain this case from the file, integrate, and make the plot. The graph should be equivalent to that depicted in the top half of Figure A1.6, page 229 in MDP.

(b) On the same graph as part (a), plot the solution to

$$dx/dt = 2x, \quad x(0) = 1, \quad 0 \leq t \leq 3.$$

(c) Erase the previous graph. Plot a new graph of the solution to part (b) using automatic scaling. Find the maximum value of $x(t)$ that the computer determined, and write this on the graph[11]. Add other caption material if you wish, and print the results.

[11] The F6≻[e] option is useful here.

Exercise $\aleph 2_{A1}$

(a) Find and graph the solution to

$$dx/dt = x, \quad x(0) = 2, \quad 0 \leq t \leq 1.$$

(b) Using a dashed line, plot on the same graph as that of (a) the solution to

$$dx/dt = -x, \quad x(0) = 2, \quad 0 \leq t \leq 1.$$

(c) Make a disk file of the above graph. Erase the screen. Then replot the graph by using the file you have just made.

Exercise $\aleph 3_{A1}$

(a) Using case "acubic" (*select* = 5, page 37) as a template, change the parameter values to form the following system

$dx/dt = y$, $dy/dt = -x$, with $x(0) = 0$, $y(0) = 1$, for $0 \leq t \leq 13$.

Graph and save the result. (Note that $d^2x/dt^2 = -x$ so that, given the initial conditions, $x = \sin t, y = \cos t$ is the exact solution.)

(b) Using (a), reproduce the top half of Figure A1.5, p228 in MDP.

(c) Erase the plot of (b), and produce a plot of both $x = \sin t$ and $y = \cos t$, using two windows, for $0 \leq t \leq 13$. Thus reproduce "at a stroke" all of Figure A1.5.

(d) Using the *select* = 5 "acubic" case, make graphs (using two windows) of $\sin 2t$ and $\cos 2t$ for $0 \leq t \leq 2\pi$.

The functions e^t, $\sin t$, and $\cos t$ can be *defined* as the (unique) solutions to the differential equations and boundary conditions that have been the subject of Exercises $\aleph 1_{A1}$— $\aleph 3_{A1}$. The various properties of these functions can be rather neatly deduced from these definitions. This approach is used, for example, in Segel (1980), appendix 1.

Exercise $\aleph 4_{A1}$

Use menu choice F7≻[c] to examine the effect of changes in integrator control parameters. Use the graphs from Exercise $\aleph 3_{A1}$ as a test case.

(a) See how large "etol" can be made without noticeable changes in accuracy. What if a 20% error is acceptable? Observe the effect that changes in "etol" have upon the speed of computation.

(b) "Play" with the other integrator control parameters and observe their effects upon speed and accuracy. Use F6≻[p] . Also play with F3≻[#].

(c) Strive for an optimal set of parameters that will minimize integration time while yielding graphs that do not differ significantly from very accurately computed graphs.

Exercises for appendix 4 of MDP

Exercise $\aleph 1_{A4}$

(a) By introducing $x = dy/dt$ (similarly to the substitution used in Exercise $\aleph 2_{A1}$), employ the "acubic" case ($select = 5$, page 37) to graph the solution to Exercise A4.1(d) on page 249 in MDP for $0 \leq t \leq 4$.

(b) Use "acubic" to graph the solution of Exercise A4.1(e) on page 249 in MDP for $0 \leq x \leq 2$.

(c) Use "acubic" as a template to graph the solution of Exercise A4.4(c) on page 249 in MDP for $0 \leq t \leq 4\pi/3$.

(d) On the same graph as (c), have the computer draw the solutions when initially $dx/dt = 3$ and when $dx/dt = 6$.

Exercises for appendix 5 of MDP

The first six Exercises can be done with the *select* = 5 case, page 37), using the files called *acubic.par* as a template and changing parameter values as appropriate.

Exercise $\aleph 1_{A5}$

Consider the system *(MDP:A5.2a)* and *(MDP:A5.2b, p250)*. Reconstruct the following figures from pages 251—253 in MDP.

(a) Figure A5.1,

(b) Figure A5.3,

(c) Figure A5.2,

(d) Figure A5.4,

(e) Figure A5.5.

Exercise $\aleph 2_{A5}$

This exercise concerns Figure A5.9 on page 258 in MDP, the vitally important summary of qualitative phase plane behavior in the neighborhood of steady–state points. Consider the perturbation equations (see the caption of Figure A5.9)

$$\frac{dx'}{dt} = ax' + by', \qquad \frac{dy'}{dt} = cx' + dy' \; . \tag{7.44}$$

(a) When $a = 1$, $b = -10$, $c = 1$, $d = -2$, which of the types of behavior that are depicted in Figure A5.9 is expected?

(b) Use the *select* = 5 equation (3.16) on page 37 to analyze equation (7.44). Use F7≻[g] to read in the *figu.par* file All parameters in equations (3.16) and (3.17) are set equal to zero in this case except for $a = a_{10}$, $b = a_{01}$, $c = b_{10}$, and $d = b_{01}$. Use the F7≻[u] option to vary the values of these four non–zero parameters.

Use F5≻[A] to draw a grid of flow arrows, with 20 arrows along the horizontal axis. Is the expected behavior consistent with this picture? Use F≻[m] to draw a typical trajectory.

(c) As shown in MDP, appendix 4, the basic solutions of the perturbation equations of (7.44) have the form $\exp(mt)$, where m satisfies *(MDP:A4.11, p246)*:

$$m^2 + \beta m + \gamma = 0, \qquad \beta \equiv -(a+d), \qquad \gamma \equiv ad - bc. \quad (7.45)$$

Find the roots of the quadratic for m, given that $\sqrt{(31)} = 5.5678$. Why should the results of part (a) have led to you to anticipate that the two roots would be complex conjugates, with the real part negative? Compare your hand calculation with what the computer provides when the steady–state is sought using F5≻[c].

Remarks:

- The computer designates the roots of the quadratic by the mathematical term "eigenvalues".
- Before employing F5≻[c] use F5≻[p] to position the cursor somewhere on the screen. Then watch the cursor move as a successive approximation procedure locates the steady–state point.
- The coordinates of the steady state will not be found exactly– but note that the results are *very* close.

(d) Erase the screen, and have the computer plot the nullclines. They should be straight lines through the origin. Why? Reset the initial point to (0.5,0.5), the value employed in the integration of part (b). Redo this integration in several time segments of duration 2, employing the F5≻[t]. Place arrowheads along this trajectory between segments by using the F5≻[a]. Move the cursor to the beginning of the trajectory and integrate *backward* for 2 time units. This can be done using F5≻[t] by replying to the question "integrate how much lcnger?" with a negative number. Plot a substantial part of at least one more trajectory, starting with the cursor near the lower right corner of the screen. Note what happens when the trajectory lies outside of the prescribed borders of the picture.

(e) Given the identification of the parameters a, b, c, d described in part (b), employ F7≻[u] to make $d = 0.5$ and $b = 0.25$. Show by hand calculations that the steady–state point is an unstable node. Check your finding with the computer. Draw the nullclines. Add a few trajectories. Hint: For unstable points, it is probably advisable to integrate backward in time, so that the trajectory will head toward the steady–state point and not off the screen.

(f) Consider the previous case with b changed to $b = -20$. Again, use hand calculations to predict the type of steady–state point and check with the computer. What about $d = -8$, $b = -12$?

(g) Verify that the case $b = 2$, $d = 1$ is a saddle point. Draw nullclines and enough trajectories so that typical saddle point behavior is clearly revealed.

Exercise $\aleph 3_{A5}$

(a) Construct several phase plane trajectories for the differential equations of *(MDP:A5.31, p263)*.

(b) Construct several phase plane trajectories for the differential equations of *(MDP:A5.32, p263)*.

(c) Display the nullclines and a few phase plane trajectories for the system *(MDP:A5.33, p266)* with the parameter values of Exercise A5.10(e).

(d) Repeat (c) for all other qualitatively different phase portraits, if any.

Exercise $\aleph 4_{A5}$

(a) Use the computer to verify your conjectures for Exercise A5.11(e), page 267 in MDP.

(b) Repeat (a) for Exercise A5.11(f).

Exercise $\aleph 5_{A5}$

Find the nullclines and a few typical phase plane trajectories for the system *(MDP:A5.36, p269)* of Exercise A5.12.

Exercise $\aleph 6_{A5}$

Use the *select* = 12 case on page 53 with the case file *lorenz.par* as a template and construct the first quadrant of the $A - B$ phase plane for the four different cases mentioned in part (d) of Exercise A5.14 on page 270 in MDP. Compare the results with Figures 8.20 and 8.21 on pages 197–198 in MDP.

Exercise $\aleph 7_{A5}$

(a) Employ case "liver" (*select* = 4) to reproduce Figure A5.12(a) and (b). The file called *liver.key* generates the solution to the latter half of this exercise.

(b) Investigate the situation wherein either or both of inequalities *(MDP:A5.20a)* and *(MDP:A5.20b)* fail to hold.

Miscellaneous Exercises

Exercise $\aleph 1_{MISC}$

Hodgkin and Huxley (1952)* received a Nobel prize for deriving equations that successfully model nerve conduction in the giant axon of the squid. A much-studied caricature of these equations, which contains many of their essential features, is associated with the names of Fitzhugh and Nagumo. For a "space-clamped" nerve[12] the Fitzhugh-Nagumo equations for the membrane potential v and a recovery variable w are

$$\frac{dv}{dt} = -v(v-1)(v-a) - w + I \tag{7.46}$$

[12]wherein spatial variation is eliminated

$$\frac{dw}{dt} = \epsilon(v - \gamma w) \tag{7.47}$$

Here ϵ and a are parameters, $\epsilon > 0$, $0 < a < 1/2$. The positive parameter I represents a constant but adjustable input current.

(a) Use the *select* = 5 case on page 37 to study this equation when $I = 0.035$, $\epsilon = 0.008$, $\gamma = 2.54$ and $a = 0.14$. Show that there is a stable steady state as well as a stable limit cycle.

(b) Repetitive firing in space-clamped squid axons can be squelched by an appropriately timed pulse of the proper magnitude (Guttman, Lewis and Rinzel, 1980)*. Baer and Erneux (1986)* showed that this "annihilation" phenomenon can be reproduced for the present case of the Fitzhugh-Nagumo equations. Try to repeat their results.

For considerable theoretical discussion of annihilation see Winfree (1980). The paper of Rinzel and Keener (1983)* provides an example of analytic work on "nerve conduction waves" governed by the Fitzhugh-Nagumo equations.

Exercise $\aleph 2_{MISC}$

Ermentrout (1984)* has added an additional excitatory variable to the Fitzhugh-Nagumo simplification of the Hodgkin-Huxley equations for axonal signal propagation (see exercise $\aleph 1_{MISC}$ above) and to the equation of Wilson and Cowan (1972)* for excitatory pyramidal cells and inhibitory interneurons. He provides an analytic argument that yields the following system to describe the essence of the behavior of the original models:

$$\frac{dx}{dt} = y - \mu x \tag{7.48}$$

$$\frac{dy}{dt} = z - \eta x \tag{7.49}$$

$$\frac{dz}{dt} = qx^2 - \gamma x \tag{7.50}$$

Here μ, η, γ, and q are positive parameters.

(a) It is asserted that a stable periodic solution exists for $2 < \gamma < 3.089$ if the other parameters are fixed at $\mu = 1$, $\eta = 2$, $q = 2$.

Check this assertion for at least one value of γ. Graph the three components as functions of time and also graph the projections of the limit cycle on the $x-z$ and $y-z$ planes.

(b) Successive period doublings are said to occur at the following approximate values of γ: 3.351, 3.410, 3.423. Again, check these assertions.

Exercise $\aleph 3_{MISC}$

Exercise $\aleph 1_{MISC}$ above deals with the Fitzhugh-Nagumo caricature of the Hodgkin-Huxley nerve conduction equations. A similar set of simple but revealing equations (proposed by Fitzhugh) are the Bonhoeffer-van der Pol equations

$$\frac{dx}{dt} = c\left(y + x - \frac{x^3}{3} + I\right)$$

$$\frac{dy}{dt} = -\frac{x - a + by}{c}$$

(a) These equations are studied in Section 2.2 of the book on neural learning models by Genis (1985)*. There it is asserted that if

$$1 - \frac{2b}{3} < a < 1, \quad 0 < b < 1, \quad b < c^2,$$

then there is a unique stable steady state when $I = 0$. Check this assertion analytically.

(b) Use the computer to check that a sufficiently large uniform depolarizing current (I sufficiently large) can induce oscillations.

Exercise $\aleph 4_{MISC}$

Chapter 2 of MDP briefly discusses period-doubling and chaos for a simple difference equation. The period-doubling "route" to random behavior has been found to be remarkably common. This exercise guides the reader to a preliminary exploration of this matter for the Lorenz differential equations (3.67)—(3.69), given on page 54 of this manual.

The system was originally derived as an extremely drastic simplification of meteorological equations, to illustrate the thesis of E. Lorenz that long term weather forecasting may be impossible – because even the slightest error in measuring conditions today may become magnified in a few days so as to destroy any predictive accuracy. The Lorenz equations have been intensively studied for the last few years. The focus of mathematical attention on this behaviorally rich yet relatively simple system is analogous to biologists' concentration on behaviorally complex yet relatively simple organisms such as *E. coli, Aplysia,* and *Dictyostelium discoideum.* The Lorenz equations do not appear to model any interesting biological situation, but their study should be of interest to biologists who wish to further their understanding of nonlinear phenomena. For additional information, consult the book of Sparrow (1982)*, from which the quantitative results given below were extracted.

(a) Use the *select* = 12 case "lorenz" on page 53, suitably modified, to show that the Lorenz equations have a periodic solution for $r = 260$[13]. Plot a projection of the limit cycle in the $x - y$ plane.

(b) Print the limit cycle graph, or sketch it, and then erase the screen. Now integrate for $r = 222$ and note the period doubling[14].

(c) Show that the period doubles again when $r = 216.2$. To make more certain that the graph repeats itself after four cycles, you may wish to sample points more frequently.

(d) Show that for $r = 213$ the solution is chaotic.

(e) Make some calculations to check the following[15]. There is a periodic orbit for (approximately) $99.98 < r < 100.795$, with a period doubling for $99.629 < r < 99.98$. Further period doublings lead to chaos. There is also a periodic orbit for $154.4 < r < 166.7$, and

[13]Hint: Employ initial conditions $x = 0$, $y = 0$, $z = 100$ and use F5≻[t], to integrate by segments until the transients have died away and the periodicity is revealed.

[14]Without changing initial conditions, employ F5≻[t] again, in order to start the new integration on the previous limit cycle and thereby minimize transients.

[15]bearing in mind the caveat at this chapter's beginning that different numerical integration schemes will lead to slightly different bifurcation values of parameters

a doubling for $148.2 < r < 154.4$. Further period doubling leads to chaos at $r = 166.3$.

Exercise $\aleph 5_{MISC}$

This is a largely analytic exercise concerning the Lorenz equations, which were also the subject of exercise $\aleph 3_{MISC}$.

(a) Show that for $0 < r < 1$ there is a single, stable, stationary point at the origin.

(b) Show that for $r > 1$ the origin becomes unstable and two new stationary points appear. For the latter, demonstrate that if perturbations are proportional to $e^{\lambda t}$, then

$$\lambda^3 + \lambda^2(\sigma + b + 1) + \lambda b(\sigma + r) + 2\sigma b(r - 1) = 0$$

(c) The classical Routh-Hurwitz criterion gives conditions that are necessary and sufficient for a polynomial to have roots with negative real parts. For the polynomial

$$\lambda^3 + a_2\lambda^2 + a_1\lambda + a_0$$

the conditions are $a_2 > 0$, $a_2a_1 > a_0$. (See for example Segel 1980, pp.171-3.) Employing this criterion, show that all roots have negative real parts if and only if

$$r < \frac{\sigma(\sigma + b + 3)}{\sigma - b - 1}$$

and therefore when $\sigma = 10$, $b = 8/3$ there are no stable steady states when $r > 2.474$.

(d) Conjecture qualitative behavior for $0 < r < 3$. Test your conjectures by simulations.

Comments

As is shown in part by the simulations requested in the previous exercise, there is a wealth of complicated behavior that is barely hinted at by the results of linear stability theory. The preceding exercise thus provides an antidote to feelings of overconfidence that a linear stability analysis plus a few simulations can reveal all the riches implied by non-linear equations. Nonetheless such an approach always provides a good start. Moreover there are many problems where the relatively simple dynamic behavior seems to be of primary biological interest.

Exercise ℵ6$_{MISC}$

(a) Find the steady states of the discrete predator–prey equations (4.16) and (4.17) and examine their stability. The stability analysis will require adapting the ideas of MDP to second order systems of discrete equations. Note that the F5≻[c] option will find critical points for the second order system.

(b) Using the results of (a), conjecture the qualitative behavior of the solution, at least in certain parameter ranges, and challenge your conjectures with simulations.

Exercise ℵ7$_{MISC}$

Tracqui et al (1985)* have analyzed the following pair of equations, which they derived in the course of modeling experiments concerning calcium metabolism:

$$\frac{dx}{dt} = a - k_1 x - (k + y^2)x + k_2 y, \qquad \frac{dy}{dt} = b + (k + y^2)x - (1 + k_2)y.$$

(a) Tracqui et al (1985)* found "tristability" for

$$k_1 = 0.5, \quad k_2 = 4, \quad b = 0.45415, \quad a = 2.6302, \quad k = 0.15244.$$

See if you can correctly conjecture the meaning of "tristability" and recover their results, for similar parameter values. Use case *acubic.par* of *select* = 5.

(b) Additional interesting behavior should be found in the neighborhood of the parameter values

$$k_1 = 0.15 \quad k_2 = 3 \quad b = 0 \quad a = 2.23 \quad k = 0.02075.$$

See what you can discover.

(c) As a project, begin a major analytic-numerical analysis of the above equations.

(d) Spatial structures are clearly important in the study of calcium involvement in bone formation. Tracqui et al (1985)* state that they are extending their work to check for the emergence of spatial structures when diffusion is added to their original equations. Try to anticipate some of their results with the aid of *select* = 20, case *cubring*.

Comments

There has been considerable effort in understanding the regulation of calcium metabolism via *compartmental analysis*. In this classical approach, linear differential equations are formulated to describe the flow of radioactive tracers between various well-mixed domains or "compartments". (In the case of calcium, typical compartments would be bone and plasma.) Chapter 3 of Rubinow (1975), for example, provides an introduction to this type of modeling.

The authors of the work that is introduced by the current exercise have shown, together with others, that in blood plasma labelled calcium oscillates with time, with a period of about one day (circadian oscillation). There is considerable evidence that the calcium oscillations are relatively independent of external environment and of hormonal status. The authors thus have proceeded under the assumption that the model must be a dynamic nonlinear system of equations that exhibits sustained oscillations. Their recent work suggests that the complicated dynamic nature of bone calcification might be the locus of the required nonlinearity — see Staub et al (1981)* and Brezillon et al (1986)*.

Exercise $\aleph 8_{MISC}$

In Chapter 6 we examined a situation wherein a limit cycle existed for certain parameter ranges while for other parameter ranges there was a single stable node. It turns out to be of considerable mathematical interest to inquire into how the limit cycle disappears under various circumstances. (Full understanding of the transition between oscillation and excitability could also lead to results of biological significance.) To obtain some insight concerning this matter, consider the following "model" problem, which contains several of the major features of *(MDP:6.15, p111)*.

$$\epsilon\,\frac{dx}{dt} = y - (\frac{x^3}{3} + \frac{x^2}{2}), \qquad \frac{dy}{dt} = -x - \alpha\,; \qquad 0 < \epsilon \ll 1.$$

(a) Find the singular point and show that it is unstable if and only if $0 < \alpha < 1$.

(b) Use case *acubic.par* (*select* $= 5$), changing parameter values as appropriate, to plot the limit cycle and neighboring trajectories, for small values of ϵ. Do this for each of a sequence of values of α that starts at $\alpha = 0.5$ and decreases through zero.

Comments

For small ϵ, in the $x - y$ (phase) plane we expect the limit cycle to consist of parts of the cubic $y = \frac{1}{3}x^3 + \frac{1}{2}x^2$ together with horizontal "jumps" from one portion of the cubic to another. (See the solution to Exercise $\aleph 20_6$ p195.) You should find that this is indeed the case, although the location of the jumps is not at all easy to predict in advance. High accuracy is needed to obtain much of the detail, which is difficult to reveal by a purely numerical investigation. See Eckhaus (1983)* for an elegant analysis of this and related problems, together with references to the earlier application to this type of problem of a new branch of mathematics called nonstandard analysis.

Exercise $\aleph 9_{MISC}$

It is one of the advantages of theoretical work that one can discern commonalities in what are superficially diverse phenomena. This exercise concerns such an instance.

Chapter 8 of MDP demonstrated how a combination of chemical reaction and diffusion can generate spatial patterns that may play an important role in morphogenesis. Segel and Jackson (1972)* pointed out that equations of similar mathematical form, but very different interpretation, might be appropriate in ecological situations and might therefore provide a mechanism for the generation of spatial inhomogeneity. Consider, for example, an uneven spatial distribution of predators and their prey. This could come about because the prey are unevenly distributed owing to geographic inhomogeneities and the predators are concentrated where the prey are most dense. But Segel and Jackson showed that such an even distribution could also arise in situations devoid of intrinsic inhomogeneity.

(a) Segel and Jackson (1972)* considered the following equations for the population densities $V(x,\bar{t})$ of victims (prey) and $E(x,\bar{t})$ of exploiters (predators), at position x and time $\bar{t}$:

$$\frac{\partial V}{\partial \bar{t}} = V(\kappa_0 + \kappa_1 V - \kappa_2 V^2) - \bar{a} V E + \mu_1 \frac{\partial^2 V}{\partial x^2}, \tag{7.51}$$

$$\frac{\partial E}{\partial \bar{t}} = \bar{b} V E - \bar{d} E - \bar{c} E^2 + \mu_2 \frac{\partial^2 E}{\partial x^2}. \tag{7.52}$$

Here, κ_0, κ_1, κ_2, $\bar{a}$, μ_1, $\bar{b}$, $\bar{c}$, and μ_2 are non-negative constants. Explain and defend the various terms. The "defense" need only consist of scenarios that show that the assumptions are conceivable, not that they are widely applicable.

(b) For simplicity, it was assumed that

$$\kappa_2 = 0, \qquad \bar{d} = 0. \tag{7.53}$$

The following dimensionless variables were introduced:

$$e = \frac{E\bar{c}}{\kappa_0}, \quad v = \frac{V\bar{b}}{\kappa_0}, \quad s = \frac{x}{(\mu_2/k_0)^{1/2}}, \quad t = \kappa_0 \bar{t}. \tag{7.54}$$

Show that the dimensionless equations are

$$\frac{\partial v}{\partial t} = (1 + \kappa v)v - aev + \delta^2 \frac{\partial^2 v}{\partial s^2} \tag{7.55}$$

$$\frac{\partial e}{\partial t} = ev - e^2 + \frac{\partial^2 e}{\partial s^2} \tag{7.56}$$

where

$$\kappa = \kappa_1/\bar{b}, \quad a = \bar{a}/\bar{c}, \quad \delta^2 = \mu_1/\mu_2. \tag{7.57}$$

(c) Show that spatial pattern spontaneously develops if

$$a = 1.5, \quad \kappa = 0.5, \quad \delta 2 = 0.01, \quad deltax = 0.5.$$

(d) Perform a stability analysis similar to that of MDP, Chapter 8, and show that a biologically relevant spatially homogeneous solution exists and is stable if

$$a > \kappa, \qquad \kappa < 1 \tag{7.58}$$

but that in an infinite domain instability will set in if

$$\delta < \delta_c \equiv a^{1/2} - (a - \kappa)^{1/2} \tag{7.59}$$

with a critical length given by

$$2\pi\delta^{1/2}(a - \kappa)^{1/4}. \tag{7.60}$$

(e) Use (d) to "predict" the results of (c).

(f) Check to see whether or not the pattern persists, at least for certain parameter ranges, if the simplifications (7.53) are relaxed.

(g) Use the *select* = 20 case *cubring* to verify and extend the analytical results of parts (a) through (f).

Exercise $\aleph 10_{MISC}$

In the supplement to Chapter 8 of MDP (pp 195–9), brief remarks are made about extensions of stability theory that take nonlinear terms into account. An example of such an extension is provided by Segel and Levin's (1976)* study of the equations (7.55) and (7.56). This study derives a typical result for this class of problems, that – in essence – if $\delta_c - \delta$ is a sufficiently small positive number then the magnitude of the deviation from steady state is proportional to $(\delta_c - \delta)^{1/2}$. Test this result by numerical simulation, bearing in mind that it will take a relatively long time to approach a new steady state when conditions are such that the homogeneous steady state is just barely unstable.

The analytic results are obtained for a continuous system in a spatially unbounded domain. Thought must be devoted to maximizing the anticipated agreement between the theory and the computations. One suggestion is to use the computed value of δ_c as a reference, not the theoretical value.

7.4 Partial answers to selected exercises, with comments

Exercise ℵ10_4 solution

In order to select appropriate scaled dimensionless variables, one must first somehow have an idea of the qualitative behavior of the solutions that are being sought. In the present case, we seek solutions to the system of equations *(MDP:4.13a)*, *(MDP:4.13b)* and *(MDP:-4.7d)*[16] for which the quasi-steady state assumption is roughly correct. In such instances, as we have discussed, we expect that the complex concentration $C(t)$ will change from its initial value C_0 to the value $\overline{C}$ of *(MDP:4.22, p58)* in a time of magnitude k^{-1}. During this period the substrate concentration $S(t)$ is expected to vary only slightly from its initial value S_0. In a second stage, $S(t)$ will decrease to zero, as substrate is transformed into product. According to *(MDP:4.32, p61)*, this will take a time of magnitude $(K_{\dot{m}} + S_0)/(k_2E_0))$, a time that is assumed to be much longer than the duration k^{-1} of the time interval wherein the complex concentration is changing rapidly.

It is reasonable to surmise that appropriate scales will differ for the initial "fast change" interval, and for the later quasi-steady interval. We shall soon see that this is indeed the case. Let us therefore examine the two intervals separately.

Since the fast-change interval lasts for a time of magnitude k^{-1}, it is natural to choose k^{-1} as the time scale for this interval. We anticipate that S varies only slightly from S_0 during the fast–change interval, so that S_0 is the natural scale for S. At first, either of two scales seem reasonable for the complex, the initial value C_0 or the final value $\overline{C}$. Of these, the latter is preferable. One reason is that the most natural value of C_0 is zero, so that on this account C_0 would be a poor choice of scale. A second reason can be inferred from the characterization *(MDP:4.21, p56)* of the time scale. To formulate this reason, we digress briefly from our particular problem to discuss some important general aspects of choosing scales for a function f.

[16] These are equations (3.20)—(3.22) on page 41 in this book.

For simplicity we *restrict ourselves to positive* f, the case of interest here. To begin with, we suppose that the maximum of f is at least somewhat bigger than its minimum. Then to a reasonable approximation *(MDP:4.21)* yields

$$\text{Time scale of } f(t) = \frac{f_{\max}}{|df/dt|_{\max}}. \tag{7.61}$$

This simplified formula is useful because *to form scaled dimensionless variables one generally employs an estimate of the maximum value of the function as its scale, and one makes the time dimensionless by dividing by the time scale.*[17] Note: equation (7.18) gives a similar albeit slightly less accurate recipe for obtaining a time scale.

Let us use F and T to denote the dimensionless counterparts of the function f and the independent variable t. Then

$$F = \frac{f}{f_{\max}}, \tag{7.62}$$

$$T = \frac{t}{f_{\max}/\left(|df/dt|_{\max}\right)}. \tag{7.63}$$

Note that equation (7.62) implies that

$$|F(T)| \leq 1. \tag{7.64}$$

Moreover, by *(MDP:4.21, p56)* and the procedure of MDP's appendix 7

$$\frac{\mathrm{d}F}{\mathrm{d}T} = \frac{f_{\max}/|df/dt|_{\max}}{f_{\max}} \frac{\mathrm{d}f}{\mathrm{d}t}.$$

That is

$$\frac{\mathrm{d}F}{\mathrm{d}T} = \frac{1}{(df/dt)_{\max}} \frac{\mathrm{d}f}{\mathrm{d}t} \tag{7.65}$$

so that

$$\left|\frac{\mathrm{d}F}{\mathrm{d}T}(T)\right| \leq 1. \tag{7.66}$$

Note the important property that *scaled dimensionless variables and their derivatives are less than or equal than unity in magnitude.* This statement should not be interpreted too literally, because of the approximate nature of our derivation and because one generally only has estimates of the appropriate scales, (For example k^{-1} is an *estimate* of

[17] Spatial variables are made dimensionless by dividing by the analogous space scale.

the duration of the interval during which the complex concentration is rapidly varying.)

It proves convenient to *define scaled dimensionless variables whose values and derivatives not only are less than unity but also never drop too far below unity.* By comparing (7.62) and (7.64), and also (7.65) and (7.66), we see that to accomplish this, *one must employ different sets of scales if the function or its derivatives decrease markedly below their maximum values.* In our example of enzyme kinetics the complex changes very rapidly for a time but later changes much more slowly—and we shall indeed employ two different scales for the two different time intervals.

We are ready to introduce scaled dimensionless variables. We shall only treat the (most important) case wherein

$$C_0 = 0, \qquad P_0 = 0. \tag{7.67}$$

We begin with the "fast change" interval. Since dependent variables are scaled by an estimate of their maxima, we introduce the dimensionless variables

$$s = S/S_0, \quad c = C/\overline{C}. \tag{7.68}$$

We have already observed that k^{-1} is the appropriate time scale, so that we define the dimensionless "fast" variable τ by

$$\tau = t/k^{-1}. \tag{7.69}$$

To estimate the amount of product at the end of the "fast" interval, we observe from *(MDP:4.7d, p53)* that k_2C is the rate of product formation. Since no product is present initially, we can estimate the amount at the end of the fast period by multiplying an estimate of the length of this period (k^{-1}) by an average rate of product formation. Since C is initially zero and has the value $\overline{C}$ at the end of the fast period, this average is $k_2\overline{C}/2$. We delete the factor 2, for we seek only rough estimates. Thus the dimensionless and scaled product variable p_i ("i" for "initial") is defined by

$$p_i = P/k_2\overline{C}k^{-1}. \tag{7.70}$$

In terms of the dimensionless variables of (7.68)—(7.70), the governing equations *(MDP:4.13a)*, *(MDP:4.13b, p54)* and *(MDP:4.7d, p53)* take the form shown in equations (3.23)—(3.25) on page 41 in the *select* = 7, *IFDIM* = 1 case.

We turn now to the time domain wherein the solution changes relatively slowly. Let us now employ T as the appropriate dimensionless and scaled time variable. From our earlier remarks concerning "long" time scale *(MDP:4.32, p61)*, we observe that the appropriate scale now is

$$\frac{K_m + S_0}{k_2 E_0} \equiv \frac{\tilde{S}_0 + 1}{k_2 \tilde{E}_0}.$$

Hence we define T by

$$T = \frac{t}{(\tilde{S}_0 + 1)/(k_2 \tilde{E}_0)}. \tag{7.71}$$

Both S_0 and $\overline{C}$ remain reasonable estimates for the maximum values of substrate and product concentration, so that the dimensionless variables (7.68) remain appropriate. But with our assumption of irreversible transformation of substrate into product (and even if we generalize our analysis by making a more realistic assumption of a weak product-to-substrate back reaction), we see that S_0 is a proper scale for a new dimensionless product variable p:

$$p = P/S_0. \tag{7.72}$$

With the present dimensionless variables the governing equations now take the form (3.30)—(3.32) on page 42

Answers for chapter 5 exercises

Exercise $\aleph 4_5$ solution

Hint: introduce dimensionless variables. Then use the *select* = 1 case "chemo" on page 30

Answers for chapter 6 exercises

Exercise $\aleph 3_6$ solution

(c) For all parameters fixed except σ, transition to relay is made between $\sigma = 3.0$ and $\sigma = 3.5$.

Exercise $\aleph 5_6$ solution

We wish the cells to be excitable, for it is these cells that are capable of being stimulated to relay. When the stimulus is applied the cells should be in a steady state and hence essentially inactive. As for the horizontal scale, it should be set to indicate approximately a single pulse, not more.

Exercise $\aleph 13_6$ solution

Here we produce the derivation of Equations (1), p. 102 of MDP. We consider the case of a cell suspension, wherein each cell is associated with a certain extracellular fluid volume V_{out}. It is assumed that at any time t, the concentration of cAMP in the extracellular volume is uniform, with value $C(t)$. The intracellular conversion of ATP (concentration $A(t)$) into cAMP (concentration $B(t)$) is assumed to take place in a narrow "reaction layer" of width d, near the cell membrane.

Assume that there is a constant input of substrate ATP into the reaction layer, at the rate of F molecules per unit time per unit area. If the area of the cell membrane is a, then approximately Fa molecules of ATP enter the reaction layer per unit time.

From equation *(MDP:4.60)* we see that (under the assumptions of chapter 4 of MDP) each adenylate cyclase enzyme molecule catalyzes the product of

$$2k\Phi(S_m, P)$$

molecules of product per unit time. The constant k gives the number of product molecules formed per unit time, per subunit. The factor 2 appears since we are assuming that the adenylate cyclase is a dimer, *i.e.* has two subunits. The factor Φ, $0 \le \Phi \le 1$, accounts for the fact that unless both enzyme and product concentrations are very high, not all the enzyme will be in a state of maximum catalytic activity.

Let E be the number of cyclase molecules per unit area of membrane, so that Ea is the total number of cyclase molecules per cell. Assume that although *(MDP:4.60)* was derived for enzyme molecules in solution, the formula may still be used in spite of the fact that the cyclase molecules are bound in the membrane[18].

Let us now consider the change during a short time interval Δt in the total number of ATP molecules in the reaction layer. We have

$$\Delta(Ada) = Fa\Delta t - 2kE\Phi a\,\Delta t. \tag{7.73}$$

Dividing by $da\,\Delta t$ and taking the limit as $\Delta t \to 0$, we obtain *(MDP:-6.1a)* if we make the definitions

$$V \equiv F/d, \qquad W \equiv 2kE/d, \tag{7.74}$$

as well as the correspondences between *(MDP:4.60)* and *(MDP:6.1d)*

$$A/K_s \equiv S_m, \qquad C/K_p \equiv P. \tag{7.75}$$

In deriving new equations, it is always wise to check dimensions. In the present case, we note that if *(MDP:6.1a)* is to be dimensionally consistent then both V and W must have dimensions of concentration/time, the dimensions of the left side of the equation[19]. Indeed, V can be regarded as the quotient of a number of molecules per unit time (Fa) divided by a volume (da). The reader should check that W also has the appropriate dimensions.

Turning our attention to cAMP, we must discuss the rate at which cAMP is secreted from the membrane. In the absence of any definitive information on this matter, we shall assume the simplest possibility, that there is random motion of the molecules that causes them to move

[18] A necessary condition for this assumption is that the "reaction layer" is sufficiently thin, *i.e.* that d is sufficiently small.

[19] Note that Φ is a dimensionless fraction.

through a pore or channel in the membrane. Suppose that in a short time Δt, every molecule of cAMP that is within distance $\delta \Delta t$ of the membrane has a certain probability θ of changing sides, *i.e.* changing from extracellular to intracellular or the reverse. Such an assumption would be justified if every molecule had a fixed average speed component v normal to the membrane, if half the molecules moved toward the membrane and half moved away from the membrane, and if the pores occupied a fraction ψ of membrane area. Then $\theta = 0.5\psi$ and $\delta = v$.

With the above simple and tentative model for motion through the membrane, we can write the following equation for the change during a time interval Δt of the number of cAMP molecules within the reaction layer:

$$\Delta(Bda) = 2kEa\Phi\Delta t - B\theta\delta\Delta t + c\theta\delta\Delta t. \tag{7.76}$$

Letting $\Delta t \to 0$, if we make the definitions

$$\theta\delta/da \equiv k_t^* \tag{7.77}$$

we obtain

$$dB/dt = W\Phi - k_t^*(B - C). \tag{7.78}$$

Turning finally to the extracellular cAMP molecules, we find that

$$\Delta(CV_{out}) = B\theta\delta\Delta t - c\theta\delta\Delta t - CV_{out}k^*\Delta t. \tag{7.79}$$

Here we have assumed that in a small time interval Δt, a fraction $k^*\Delta t$ of cAMP molecules will be hydrolyzed (destroyed) by the enzyme phosphodiesterase. In other words, k^* is the probability per unit time that a given cAMP molecule will "successfully" collide with some phosphodiesterase molecule. (Note that k^* will be proportional to the extracellular cAMP concentration. We are in essence using *(MDP:4.24)* under the assumption that the extracellular cAMP concentration is small compared to the Michaelis constant K_m of the phosphodiesterase.)

From the equation just obtained, a passage to the limit yields

$$dC/dt = k_t^* h{-}1(B - C) - k^*C \tag{7.80}$$

if

$$h{-}1 = dA/V_{out}, \tag{7.81}$$

i.e. if h is the ratio of the extracellular fluid volume per cell to the volume of the reaction layer. (Note the difference with the definition of h given on p. 103 of MDP.)

Equations *(MDP:6.1b)* and *(MDP:6.1c)* are finally recovered if we make the approximation $B - C \approx B$, as explained on p. 103 of MDP.

Exercise $\aleph 19_6$ solution

Consider the problem of estimating the ATP (α or A) time scale. This can be done by focusing attention either on relay or on oscillation, both of which exhibit similar time scales. It seems easier to consider relay; it is triggered at a specific time, giving a natural starting-point for our analysis.

Let us estimate the time scale for the first, decreasing, phase of ATP change. We shall do this by estimating how long it takes for the dimensionless ATP concentration α to be reduced from its steady state value $\overline{\alpha}$ to zero, using the maximum rate of conversion of substrate ATP into product cAMP. We shall employ an overestimate of the actual change, as well as an overestimate of the rate of change. Presumably there will be some cancellation of error.

Equation *(MDP:6.9, p105)* gives an explicit expression for $\overline{\alpha}$ in terms of the parameters of our problem, but this formula is too complicated for our purposes. In attempting to simplify it, we shall restrict ourselves to the situation (true for our simulations) where

$$\sigma \gg 1, \quad L \gg (1+\overline{\gamma})^2 \quad i.e. \; L \gg \left(1+(q/hk)\right)^2. \tag{7.82}$$

Then the quadratic *(MDP:6.7, p104)* can be approximated by

$$\sigma\overline{\alpha}^2 + \sigma\overline{\alpha} - (1+\overline{\gamma})^{-2}L = 0.$$

We anticipate from Figures 6.5 and 6.6 of MDP that $\overline{\alpha} \gg 1$. Thus the term $\sigma\overline{\alpha}$ should be small compared to $\sigma\overline{\alpha}^2$, and we obtain

$$\overline{\alpha} \approx \left(\frac{L}{\sigma}\right)^{1/2} \frac{1}{1+\overline{\gamma}} \equiv \left(\frac{L}{\sigma}\right)^{1/2} \frac{1}{1+(q/hk)}. \tag{7.83}$$

(This result can also be obtained directly by approximating *(MDP:-6.9)*.) As one check on the estimate (7.83) we find, employing the parameters of Figure 6.6 (MDP, p108), that our estimate gives $\overline{\alpha} \approx 91.3$, which is an excellent approximation to the correct value $\overline{\alpha} = 92.4$. Indeed, in estimating orders of magnitudes we really do not need this good an approximation. Since $\overline{\gamma} = 1$ for the parameter of Figure 6.6, we could get a decent estimate with the simpler approximation $\overline{\alpha} \approx (L/\sigma)^{1/2}$. But to achieve more comprehensive results, we shall retain the more accurate formula.

Given the steady state value of A, $\overline{A} = \overline{\alpha} K_s$, we wish to estimate how long it will take for this amount of ATP to be exhausted by maximum operation of the reaction for ATP-cAMP conversion. The maximum rate of this reaction is given by the parameter W. Thus we estimate the time scale for the first phase of the ATP change to be

$$\left(\frac{L}{\sigma}\right)^{1/2} \frac{1}{1+(q/hk)} \frac{K_s}{W}. \tag{7.84}$$

With the parameters of MDP Figure 6.6, this estimate gives 76 sec, which compares quite favorably with the computer simulations.

We estimate the duration of the ATP replenishment phase by calculating how long it takes the ATP concentration to build up from zero to $\overline{\alpha} \equiv \alpha K_S$. We take the maximum value V of the build-up rate, ignoring some continued ATP-cAMP conversion. The result is

$$\left(\frac{L}{\sigma}\right)^{1/2} \frac{1}{1+(q/hk)} \frac{K_s}{V}. \tag{7.85}$$

With the parameters of Figure 6.6 this estimate yields the satisfactory value of 912 secs.

We are now in a position to reap the fruits of our labors by collecting and summarizing our results.

(i) We can estimate the period of the cAMP oscillation and the total length of a relay pulse, including the recovery period, by

$$\left(\frac{L}{\sigma}\right)^{1/2} \frac{1}{1+(q/hk)} \frac{K_s}{V} \left(1+\frac{V}{W}\right). \tag{7.86}$$

In the cases we have studied, $V/W \approx 0.1$ so that the factor $1 + (V/W)$ can be ignored. The fact that $V \ll W$ is primarily responsible for the spike–like nature of the cAMP pulse in Figures 6.5 and 6.6 of MDP and for the corresponding relatively slow traverse of the vertical left portion of the limit cycle in Figure 6.12 (MDP, p116).

(ii) For a quasi–steady–state assumption on the extracellular cAMP concentration C to be valid, the time scale for C change must be short compared to the faster time scale for ATP change:

$$\max\left((k_t^*)^{-1}, (k^*)^{-1}\right) \ll \left(\frac{L}{\sigma}\right)^{1/2} \frac{1}{1+(q/hk)} \frac{K_s}{W}. \tag{7.87}$$

(iii) A quasi-steady state assumption on B requires that the two time scales for changes in B be small compared to the corresponding two time scales for changes in A, in the two phases of ATP diminution and replenishment. This yields

$$q^{-1} \ll 1, \quad (k_t^*)^{-1} \ll \left(\frac{L}{\sigma}\right)^{1/2} \frac{1}{1+(q/hk)} \frac{K_s}{V}. \tag{7.88}$$

All the above estimates were made under the supposition that σ and L are large in the sense of (7.82).

(iv) Conditions in addition to those of (7.87) and (7.88) could be necessary to insure the validity of the quasi–steady–state approximations. But initial conditions are not generally important for the present problem, so that we need not concern ourselves with a discussion of the type that led to condition (6.4) for the validity of the classical quasi–steady–state approximation.

Here are some general implications of our careful analysis of the particular problem upon which we have been concentrating our attention.

(i) Estimates of magnitudes and scales generally require certain restrictions on the parameters of the problem. The estimates might be very different under different restrictions.

(ii) Order of magnitude estimates usually permit deeper understanding of the process in question. Consider for example the approximation $\overline{\alpha} \approx (L/\sigma)^{1/2}$ for the dimensionless steady state of ATP that was discussed under equation (7.83). This estimate for $\overline{\alpha}$ follows from *(MDP:-6.3)* if $\phi \approx \alpha^2/L$. Among the implications of this approximation of ϕ is that, *for the parameter ranges under consideration,* at steady state there is not sufficient accumulation of extracellular cAMP to influence the catalytic rate, and moreover that ATP is not in the range where substrate saturation effects begin to make themselves felt.

(iii) Oscillations or pulses of a variable can often be analyzed by breaking their time course into two phases, one in which the dominant process is depletion of the variable and the second in which the dominant process is replenishment. There may be more than one process that contributes to replenishment, say, but often a single process will be the major one and others can be ignored. If two processes are of approximately equal importance, either can be examined to obtain a time scale.

(iv) Consider two coupled variables wherein change in the first is governed by processes that are intrinsically faster than the processes governing change in the second. Because the first variable will be in a quasi-steady state (perhaps after a transition period) the entire system will be observed to vary on the slower time scale that governs the second variable.

(v) Estimates of magnitudes and scales in terms of problem parameters is generally quite a difficult process. For serious problems the estimates are hardly conceivable without the guidance of computer simulations. The simulations give us indispensable initial qualitative clues as to the relative magnitudes of various variables and their relative rates of change. Later they allow us to check whether proposed parameter combinations do, as hoped, provide quantitative estimates for these magnitudes. Checks are carried out by examining simulation results for one or more particular sets of parameter values.

(vi) "Scaling" estimates of the type we have made provide the indispensable first steps in perturbation methods that can often provide relatively simple approximate solutions to complex equations. See Lin and Segel (1974), part B.

Exercise $\aleph 20_6$ solution

Consider the phase plane in Figure 6.12, MDP page 116, and assume that β generally changes much faster than α, *i.e.* $|d\beta/dt| \gg |d\alpha/dt|$. Suppose that at time $t = 0$ the state point is located, as shown, in the upper top left corner of the diagram. Since β changes rapidly compared to α, there will be almost no change in α as the state point proceeds nearly horizontally. Motion will be to the right since $d\alpha/dt > 0$. The rapid near-horizontal motion will cease when the state point approaches the nullcline wherein $d\beta/dt = 0$. On this nullcline, in contradistinction into the general situation, β can be expressed in terms of α by the quasi-steady state relationship obtained from $d\beta/dt = 0$; α will change because $d\alpha/dt \neq 0$. A motion down the nullcline will ensue (since $d\alpha/dt < 0$) until this is no longer possible, at the minimum of the nullcline $d\beta/dt = 0$. Here the only alternative is to "fall off" the nullcline $d\beta/dt = 0$ and again to move rapidly, nearly horizontally, this time to the left, until another branch of the nullcline $d\beta/dt$ is reached. This almost vertical branch is not shown in Figure 6.12, but is depicted in Figure 6.10. Again $d\beta/dt = 0$ but $d\alpha/dt \neq 0$. In the domain in question, $d\alpha/dt > 0$. Thus the state point will climb up the near-vertical limb of the $d\beta/dt = 0$ nullcline, relatively slowly compared to the rapid horizontal movement that occurs except near the nullcline where $d\beta/dt = 0$.

When its climb is finished, at the maximum of the $d\beta/dt = 0$ nullcline, the state point is near the sole steady state point. If this point is stable, as in the relay case of Figure 6.13, the trajectory will approach the steady state. If not, the state point is condemned to infinite repetitions of its excursion in the phase plane, bouncing back and forth between arms of the nullcline $d\beta/dt = 0$. (Note: Such rapid bouncing between nullclines, interspersed with slow travel along nullclines, is a general feature of phase portraits where one variable generally changes very rapidly compared with the other. In the periodic case, as in Figure 6.5, *relaxation oscillations* result. As opposed to the regular rise and fall of sinusoidal oscillations, here there is a period of rapid change or "firing", followed by a period of slow recovery.)

Exercise $\aleph 22_6$ solution

(c) If the initial point is a steady state the trajectory will not move off of it and the graph of the trajectory will be a point, which is probably invisible. If the initial point is very near an unstable steady state (and you almost certainly cannot start *exactly* at a steady state because of round off error) then the state-point will move, but very slowly at the beginning.

(d) Evidently the limit cycle becomes much smaller when there is a shift from $n = 2$ to $n = 4$. Change the scale of the graph.

Answers for chapter 7 exercises

Exercise $\aleph 1_7$ solution

According to the rule of thumb, to diffuse a half-circumference $L = 100\mu m = 10^{-2}$ cm, it should take $(10^{-4}\text{cm}^2)/(10^{-5}\text{cm}^2\text{sec}^{-1}) = 10\text{sec}$. To check this with a ring of $M = 4$ intervals of width δ we must take $2\delta = L$, *i.e.* $\delta = 0.5 \times 10^{-2}$. For $M = 8$ we must use $\delta = 0.25 \times 10^{-2}$.

For both cases put 8 units of chemical in cell 1, the "north pole". For $M = 4$ the south pole is cell 3. Eventually the chemical should be uniformly distributed, with 2 units in each cell. Thus, for a in particular, the scaling is correct: $0 \leq a_3 \leq 2$. For $M = 8$, cell 5 is the south pole. Eventually there should be a single unit of chemical in each cell, so that here $0 \leq a_5 \leq 1$. A good measure of how long it takes for a "substantial" quantity of chemical to arrive at the south pole is to check the time it takes for the concentration there to rise from a negligible value to half its final value.

The chemical h is governed by all the same conditions as govern a except that the half-circumference of the ring is twice as long, so that δ in each case must be twice as large.

It is left to the reader to carry out the simulations and to comment on how the results conform to the rule of thumb.

As to whether $M = 4$ or $M = 8$ should in principle yield results that are closer to the rule of thumb, it must be borne in mind that (7.26) is a better approximation of (7.22) if the subinterval length δ is shorter. For a given length L, therefore, when $M = 8$ our numerical results should be closer to the solution of the differential equation, and hence to the rule of thumb.

Exercise $\aleph 3_7$ solution

The permeability k_a is the only dimensional parameter in the problem. It has dimension $(\text{time})^{-1}$. On these grounds alone, the diffusion time should be proportional to k_a^{-1}. This reasoning is reinforced by solving the following problem for two adjacent cells.

$$\frac{da_1}{dt} = k_a(a_2 - a_1), \quad \frac{da_2}{dt} = k_a(a_1 - a_2), \quad a_1(0) = 2, \ a_2(0) = 0.$$

We have

$$\frac{d\,(a_1 + a_2)}{dt} = 0 \Longleftrightarrow a_1(t) + a_2(t) = \ \textit{constant} \ = a_1(0) + a_2(0) = 2.$$

Therefore

$$\frac{da_1}{dt} = k_a(2 - a_1 - a_1) = 2k_a - 2k_a a_1.$$

This equation is of the same form as *(MDP:4:15, p55)*. Using the method of the supplement on p73-4 of MDP we find that the solution of the above equation that satisfies the initial condition is

$$a_1(t) = 1 - \exp^{-2k_a t}.$$

This gives a time scale of magnitude k_a^{-1}.

To see the role of the cell number M, replace L by $M\delta$ in the usual rule of thumb. This gives $T = M^2\delta^2/D$ or $T = M^2/k_a$, a conjecture that should be challenged by simulation studies.

Answers for chapter 8 exercises

Exercise $\aleph 2_8$ solution

(b) The predictions of linear theory are, as anticipated, more accurate when Q is relatively close to its critical value ($Q = 0.622$) for instability. In these “weakly unstable” situations the oscillations are rather sinusoidal in form, as assumed in the linear theory. By contrast, when Q becomes small the graphs assume a pulse-like character.

Exercise $\aleph 5_8$ solution

1. (b) A variety of nondimensionalizations *can* reduce the number of parameters to the same minimum number. Here we chose to take the same scaling for activator and inhibitor concentrations ($a_0 = h_0$) in order to have independently variable dimensionless activator and inhibitor strengths. The price we paid was an increase by one in the number of dimensionless parameters.

References

References originally cited in MDP (*p291–6*) are not included in this list. Found below are new references. These are distinguishable in the text of this manual by a star following the year of publication, as in Bernhard (1968)*.

Baer, S.M., and Erneux, T. (1986). Singular Hopf bifurcation to relaxation oscillations. SIAM J. Appl. Math (in the press).

Bernhard, S.A. (1968). *The Structure and Function of Enzymes.* New York: Benjamin.

Brézillon, P., Tracqui, Ph., Milhaud, G., Perault-Staub, A.M. and Staub, J-F. (1986). Calcium metabolism in rat: II. Nonsteady state, nonlinear model; formalism and procedures. Preprint: UA 163, CHU Saint-Antoine, 27 rue Chaligny, 75012 Paris, France.

Burden, R.L., and Faires, J.D., (1985), *Numerical Analysis,* third edition, Sections 5.6 and 5.7, Boston: Prindle, Weber and Schmidt.

Byrne, G. and Cox, E.C. (1986) Spatial patterning in *Polysphondylium*: monoclonal antibodies specific for whorl preparations. *Developmental Biology*, **117**, 442–455.

Decroly, O., and Goldbeter, A. (1982). Birhythmicity, chaos, and other patterns of temporal self-organization in a multiply regulated biochemical system. *Proc. Nat. Acad. Sci. USA*, 79:6917–21.

Decroly, O., and Goldbeter, A. (1984). Multiple periodic regimes and final state sensitivity in a biochemical system. *Physics Letters*, 105:259–62.

Dixon, M., and Webb, E.C. (1958). *Enzymes.* London: Longman's, Green & Co.

Eckhaus, W. (1983) Relaxation oscillations including a standard chase on French ducks. In: *Asymptotic Analysis II*, edited by F. Verhulst (Lecture Notes in Mathematics No. 985), pp. 449–94, Berlin: Springer-Verlag.

Ermentrout, G.B. (1985). The behavior of rings of coupled oscillators. *J. Math. Biol.* **23**:55–74.

Ermentrout, G.B. (1984). Period doubling and possible chaos in neural models. *SIAM J. Appl. Math* 44:80–95.

Fersht, A. (1985). *Enzyme structure and mechanism* , second edition, New York: W. H. Freeman.

Genis, C.T. (1985). *Temporal-Pattern Learning in Neural Models*. Berlin: Springer-Verlag.

Goldbeter, A., Martiel, J.L., and Decroly, O. (1984). From excitability and oscillations to birhythmicity and chaos in biochemical systems. In: *Dynamics of Biochemical Systems*, edited by J. Ricard and A. Cornish-Bowden, 173–212.

Guttman, R., Lewis, S., and Rinzel, J. (1980). Control of repetitive firing in squid axon membrane as a model for a neuron oscillator. *J. Physiol.* 305:377–95.

Grutsch, J., and Robertson, A. (1978). The cAMP signal from *Dictyostelium discoideum* amoebae. *Devel. Biol.* 66:285–93.

Hammes, G.G., and Schimmel, P.R. (1970). Rapid reactions and transient states. pp. 67–114 In: *The Enzymes* (Vol. 2), edited by P.D. Boyer, New York: Academic Press.

Hassel, M., and Varley, G.C. (1969), New inductive population model for insect parasites and its bearing on population control, *Nature* 223:1133–7.

Hodgkin, A.L., and Huxley, A.F. (1952). A quantitative description of membrane current and its application to conduction and excitation in nerve. *J. Physiol.* 117:500–44.

Jakobson, M.V. (1981). Absolutely continuous invariant measures for one parameter families of one dimensional maps. *Commun. Math. Phys.* 81:39–88.

Jensen, M.H., and Procaccia, I. (1985). Chaos *via* quasiperiodicity: universal scaling laws in the chaotic regime. *Phys. Rev. A.*, **32**:1225-8.

Kantz, H., and Grassberger, P. (1985). Repellers, semi-attractors, and long-lived chaotic transients. *Physica D*. 17:75–86.

Knuth, D. (1984) *The TEXbook*, Addison–Wesley Publishing Company.

Lamport, L. (1986) LATEX: *A Document Preparation System*, Addison–Wesley Publishing Company.

Li, T-Y and Yorke, J.A. (1975) Period three implies chaos, Amer. Math. Monthly 82, 985-92.

Markus, M., and Hess, B. (1984). Transitions between oscillatory modes in a glycolytic model system. *Proc. Nat. Acad. Sci. USA*, 4394–8.

Markus, M., Kuschmitz, D., and Hess, B. (1984). Chaotic dynamics in yeast glycolysis under periodic substrate input flux. *FEBS Letters* 172:235–8.

Martiel, J.L., and Goldbeter, A. (1984). Oscillations et relais des signaux d'AMP cyclique chez *Dictyostelium discoideum*: analyse d'un modèle fondé sur la modification du récepteur pour l'AMP cyclique. *Compte Rendu Acad. Sci. Paris*, 298:549–52.

Nussc, H.E. (1984). Complicated dynamical behavior in discrete population models. *Niew Archief Voor Wiskunde*, 2:43–81.

Odell, G. M. (1980). Qualitative theory of systems of ordinary differential equations including the use of the Hopf bifurcation theorem. appendix 5 in *Mathematical Models in Molecular and Cellular Biology*, edited by L. A. Segel, Cambridge University Press.

Phillips, M.B., and Odell, G.M. (1984). An algorithm for locating and displaying the intersection of two arbitrary surfaces. *IEEE Computer Graphics and Applications* 4:48–58.

Rinzel, J., and Keener, J.P. (1983). Hopf bifurcation to repetitive activity in nerve. *SIAM J. Appl. Math.* 43:907–22.

Robertson, A., and Drage, D. (1975). Stimulation of late interphase *Dictyostelium discoideum* by amoebae with an external cyclic AMP signal. *Biophys. J.* 15:765–75.

Schauer, M., and Heinrich, R. (1979). Analysis of the quasi–steady–state approximation for an enzymatic one-substrate reaction. *J. Theoret. Biol.* 79:425–42.

Schwarz, I.B. (1985). Multiply stable recurrent outbreaks and predictability in seasonally forced nonlinear epidemic models. *J. Math. Biol.* 21:347–61.

Segel, L., Goldbeter, A., Devreotes, P., and Knox, B. (1985). A model for sensory response and exact adaptation mediated by receptor modification. In: *Sensing and Response in Microorganisms*; edited by M. Eisenbach and M. Balaban, pp. 175–183. Amsterdam: Elsevier.

Segel, L., Goldbeter, A., Devreotes, P., and Knox, B. (1986). A mechanism for exact adaptation based on receptor modification. *J. Theoret. Biol.*, in press.

Segel, L.A., and Jackson, J.L. (1972). Dissipative structure: an explanation and an ecological example. *J. of Theoret. Biol.*, 37:545–559.

Segel, L.A., and Levin, S.A. (1976). Applications of nonlinear stability theory to the study of the effects of dispersion on predator-prey interactions. In: *Selected Topics in Statistical Mechanics and Biophysics*; edited by R. Piccirelli, pp. 123–152. American Institute of Physics Symposium 27.

Sparrow, C. (1982). *The Lorenz Equations: Bifurcations, Chaos and Strange Attractors*. New York: Springer-Verlag.

Staub, J-F., Brezillon, P., Perault-Staub, A.M., and Milhaud, G. (1981) Nonlinear modeling of calcium metabolism – First attempt in the calcium deficient rat. *Trans. Inst. Meas. Contr.* 3(2):89–97.

Thompson, J.M.T. and Thompson, R.J. (1980). Numerical experiments with a strange attractor. *Bull. Inst. Maths. and its Appl.* 16:150–4.

Tracqui, Ph., Brézillon, P., Staub, J-F., Perault-Staub, A.M. and Milhaud, G. (1985). Diversité et complexité des comportements dynamiques temporels d'un modèle autocatalytique simple. *C.R. Acad. Sc. Paris*, 300:253–8.

Venieratos, D., and Goldbeter, A. (1979). Allosteric oscillatory enzymes: influence of the number of protomers on metabolic periodicities. *Biochemie* 61:1247–56.

Wilson, H.R. and Cowan, J.D. (1972). Excitatory and inhibitory interactions in localized populations of model neurons. Biophys. J. 12, 1-24.

Corrections and minor changes to MDP

In the following list "p3,7" means page 3, line 7. "p2,-6" means page 2, 6 lines up from the bottom.

- Frontispiece: invert lower photograph. In caption, delete "See Chapter 6 for more details". Add at end "See Chapter 6 for an introduction to modeling chemoattractant secretion in slime mold.

- p32, -4: The text of (d) should be "Show that when $r < 1$ (when there is no positive steady state) $x = 0$ is stable, and is otherwise unstable."

- p44, Fig. 3.5, lower left quadrant: should be "$\Phi_{Aa} < \Phi_{aa}$, $\Phi_{Aa} < \Phi_{AA}$".

- p47, Eq. (28): "$\frac{q_n}{q_0}$" should be "$\frac{q_0}{q_n}$".

- p49, problem 5

  - Part (a) should read "Show that in this case (where allele a is a recessive lethal)". In equation (29) replace "$p_{n+1} = 0$" by "p_{n+1} undefined". Two lines below equation (29), delete " $p = 0$ ".

  - Part (b). Delete the first two sentences. Replace "Thus show" by "Show".

  - Part (d) should read as follows. "Note that in the case under consideration, in contrast with the general situation when L is not equal to zero, the absence of all a–alleles cannot be regarded as a steady state. Nonetheless it is instructive to approximate (29) when p_n is small. What is the result of the calculation? Does it make biological sense? Given the results of the various calculations, conjecture the overall qualitative behavior."

- p65, first line under Eq. (44): "(38)" should be "(36), (38)".

- p66, Fig. 4.6, 2nd line of caption: "concavity" should be "convexity".

- p66, last paragraph: Second sentence should read "But (48) indicates that the increase in binding accelerates as ligand...".

- p67, Fig. 4.7: The rightmost of the two downward arrows should be labelled "2k", not "k".

- p70, Exercise 2. The displayed equations (on lines -17, -12, -8, -6) should be changed to the following.

$$\frac{dx}{dt} = -3x + y, \quad \frac{dy}{dt} = -100(y - 2x), \quad y(0) = 0, \quad x(0) = 1.$$

$$x \approx \exp(-t), \quad y \approx 2\exp(-t)$$

$$y = Ae^{-102.02t} + Be^{-0.98t}, \quad x = 0.01Ae^{-102.02t} + 0.495Be^{-0.98t}.$$

$$x = 0.02e^{-102.02t} + 0.98e^{-0.98t}, \quad y = -1.98e^{-102.02t} + 1.98e^{-0.98t}$$

- p71, second line of Ex. 8: "convex" should be "concave".

- p72, -2: "concave" should be "convex".

- p73, 4: "concavity" should be "convexity".

- p99, -16: change "interplay" into "operation".

- p125, Fig. 6.15: Add 1.4 from the coordinates given on both the ordinate and abscissa. Also, correct caption to read "$k' = k_1kS/V$" (line 2) and "$k_t = 10$" (line 3).

- p128, line 2 of Ex. 2(a): After "$\sigma > 1$", add ", or else the roots are complex."

- p141, last line: "sin" should be "cos".

- p164, Fig 8.7: add at end of caption: "Total length divided into 40 segments."

- p232, -6: "$f(x)$" should be "$f(a)$".

- p237, Eq. (4): "a_0N_0" should be "$a_0N_0 = 0$".

- p248, two lines under Eq. (20): delete "=0".

- p254, Eq. (12): second equation should be "$y|'(t) = y(t) - \overline{y}$".

- p256, last paragraph: Replace the second and third sentences by "In particular, the reader should check that in (17c), $m_1 > 0$ and $m_2 > 0$."
- p265, Fig. A5.12(b). The upper heavy dot on the s axis should be deleted.
- p267, -10: should be "$d\overline{y}/dt$" and "(hy^*)".
- 267, -8: "$\left(a - \frac{be}{h}\right)$" should be "$\left(\frac{ah}{c} - \frac{be}{c}\right)$".
- p269, one line above (38a,b): "$s(t)$" should be "$S(t)$".
- p270, line 4 of Ex. 13(d): "$p > qS$" should be "$p > q\overline{S}$".
- p272, Eq. (3): "$(b_2 + ib_2)$" should be "$(a_2 + ib_2)$".

Very minor changes

- p xvi:-4: "Garry", not "Gary".
- p26, Fig. 2.8: The point $x_0 = 0.9$ should be graphed and "(not shown)" deleted from the caption.
- p42,-9: "aA" should be "Aa".
- p44, Fig. 3.5: In upper right quadrant, "aA" should be "Aa".
- p73, Eq. (2): "$=$" should be "$\equiv$".
- p98, 11: Change "(PDE), which" to "(PDE). This".
- p104, Eq. (6): add period at end of equation.
- p163, Fig. 8.6; 164, Fig. 8.7; 172, Fig. 8.10: <u>all</u> parameters should have "prime".
- p198,-1 "Shuster" should be "Schuster".
- p203, Fig. 9.4: Extend lines so that they meet open circles.
- p204, 3: "normal" should be "normally".
- p228: The graph of sin x should not have a discontinuity.

- p234, Eq. (8b): partially mutilated beginning should be "$f_x(a,b)$".
- p245, Eq. (5): "α_{n-1}" should be "$\alpha_{n-1}(t)$".
- p252, third line under (7): "interesected" should be "intersected".
- p292, 10: "Shuster" should be "Schuster".
- p293, 12: "Dictyostelium" should be in italic.

Completion of references (p291-6)

- Harris et al (1983) should be "Harris,...(1984). Generation...model. J.Embryol. Exp. Morph., 80:1-20."
- "Kemner, W. (1983)..." should be "Kemner, W. (1984) A model of head regeneration in *Hydra. Differentiation*, 26:83-90."
- Oster, Murray, and Harris (1984). Change "(in press)" to "78, 83-125."
- Oster and Odell (1984). Change ".Amsterdam:North Holland" to "L. Campbell, and P. Channell. *Physica D. Nonlinear Phenomena.*, 12:33-50."
- Because of the above, the following changes should be made.
    - p183,-7: Change "(1983)" to "(1984)".
    - p220,-13: Change "(1983)" to "(1984)".

Index